Writings on Medicine

forms of living

Stefanos Geroulanos and Todd Meyers, *series editors*

Writings on Medicine

Georges Canguilhem

Translated and with an Introduction by
Stefanos Geroulanos and Todd Meyers

FORDHAM UNIVERSITY PRESS

NEW YORK 2012

Fordham University Press has no responsibility for the persistence or accuracy of URLs for external or third-party Internet websites referred to in this publication and does not guarantee that any content on such websites is, or will remain, accurate or appropriate.

Fordham University Press also publishes its books in a variety of electronic formats. Some content that appears in print may not be available in electronic books.

Writings on Medicine was first published in French as *Écrits sur la médecine*: © PUF, 1989, for the chapter "Les maladies" ("Diseases"); © Sables, 1990, for "La santé, concept vulgaire et philosophique" ("Health: Popular Concept and Philosophical Question"); and © Éditions du Seuil, 2002, for the remaining texts and the assembly of the volume.

The first quotation on the back cover of this book is reprinted from "What Is Health? The Ability to Adapt," *The Lancet*, vol. 373, no. 9666 (2009): 781, copyright 2009, with permission from Elsevier.

Library of Congress Cataloging-in-Publication Data
Canguilhem, Georges, 1904-1995.
[Ecrits sur la médecine. English]
Writings on medicine / Georges Canguilhem ; translated and with an introduction by Stefanos Geroulanos and Todd Meyers.—1st ed.
    p. cm.
Includes bibliographical references (p.   ) and index.
ISBN 978-0-8232-3431-8 (cloth : alk. paper)
ISBN 978-0-8232-3432-5 (pbk. : alk. paper)
  1. Medicine—Philosophy.   2. Medicine—History.   I. Title.
R723.C36513   2012
610—dc23
                                                2012009732

14 13 12 5 4 3 2 1
First edition

CONTENTS

ACKNOWLEDGMENTS

We would like to acknowledge the undertaking assumed by Armand Zaloszyc in organizing the original volume by Éditions du Seuil. Thanks go to Warren Breckman, Michael Carhart, and Allan Megill for their editorial interventions to an earlier draft of the introduction and to Helen Tartar and Bud Bynack at Fordham University Press for their editorial work and support throughout. Henning Schmidgen's German-language translation of three of the essays, *Gesundheit—eine Frage der Philosophie*, was a great resource. Claudio Lomnitz, then editor of *Public Culture*, was generous with both his enthusiasm and his editorial guidance on an earlier published version of Canguilhem's "Health," for which we are grateful. François Delaporte graciously offered technical (and practical) advice. Lastly, we would like to acknowledge the librarians at the Archives de Georges Canguilhem at the Centre d'Archives de Philosophie, d'Histoire et d'Édition des Sciences (CAPHES) of the École Normale Supérieure, the Max Planck Institute for the History of Science, and the Wayne State University Purdy Kresge Library for their assistance.

Writings on Medicine

Georges Canguilhem's Critique of Medical Reason

Stefanos Geroulanos and Todd Meyers

I.

At the time of his death in 1995, Georges Canguilhem was a highly respected epistemologist and historian of biology and medicine. He was known for having extended and transformed traditions set by Gaston Bachelard and Henri Bergson, and as an influential figure for generations of scholars, including Michel Foucault, François Dagognet, Louis Althusser, Pierre Bourdieu, Dominique Lecourt, Gilbert Simondon, and Gilles Deleuze. At different stages of his life, he was in conversation with important contemporaries, among them François Jacob, Jean Hyppolite, Maurice Merleau-Ponty, Jean Cavaillès, Kurt Goldstein, and René Leriche. He spearheaded both a radical undermining of scientific positivism and a retheorization of central categories of biology, medicine, and psychology in a period marked by major advances in these fields.[1]

Most of the work that placed him in such an exceptional position is to be found in five books published over the course of half a century: *Essai sur quelques problèmes concernant le normal et le pathologique* (*The Normal and the Pathological*, 1943; second, modified edition, 1966),[2] *La formation du concept de réflexe aux XVIIᵉ et XVIIIᵉ siècles* (*The Formation of the Concept of Reflex in the Seventeenth and Eighteenth Centuries*, 1955),[3] *La connaissance de la vie* (*Knowledge of Life*, 1952; second, modified edition, 1965),[4] *Idéologie et ratio-nalité dans l'histoire des sciences de la vie* (*Ideology and Rationality in the History of the Life Sciences*, 1977),[5] and *Études d'histoire et de philosophie des sciences du vivant et de la vie* (*Studies in the History and Philosophy of the Sciences of the Living and of Life*, 1983).[6] Strictly speaking, only the first two are books: the last three are collections of essays and talks, although *Knowledge of Life* and *Ideology and Rationality* each engages with a single problem and has rather clear aims.

Canguilhem is best known for his first book, *The Normal and the Patholog-ical*, which came at a crucial point in the history of medicine and which articulated a concern that had been vaguely felt across much of medical thought: that medicine, biology, and physiology rely on formal and statisti-cal norms that hamper, rather than aid, not only diagnosis and treatment, but also understanding of the individual patient's relation to society and to medical intervention. Canguilhem worked from a perspective directly influenced by the surgeon René Leriche and by the German neurologist and psychiatrist Kurt Goldstein, both of whom helped him to question the claim—which in France dated back to François-Joseph-Victor Broussais and Auguste Comte—that disease and the pathological condition more gener-ally are nothing more than modifications of the normal condition.

Canguilhem pointed out the normative and norm-producing effects that "the normal" had possessed ever since Comte: the norm was held simulta-neously to be both identical to the normal and the only norm. Conse-quently, there could be no norms specific to disease—disease could be only an aberration. Thanks to these ideas, an elaborate language emerged aimed at helping the physician see signs as symptoms—thus bypassing the individ-ual patient, for whom disease is a specific and qualitatively heterogeneous experience.

Goldstein's physiology allowed Canguilhem to suggest that pathology is much more complicated than that—in particular because Goldstein had demonstrated in the 1920s and 1930s that an organism attempts to compen-sate for damage done to certain functions and that the diseased body tends to obey different rules than the normal body, but is not, for all that, norm-less, except in catastrophic situations.[7] "Disease creates a shrunken milieu and is a narrowed mode of life, but it is also, for the individual patient, a new life, characterized by new physiological constants and new mechanisms."[8]

From Leriche, Canguilhem took up and recast the claim that "health is life lived in the silence of the organs," while disease is what "irritates men in the normal course of their lives and work, and above all what makes them suffer."[9] Canguilhem thus attended to the complexity of pathological experience in a way that was at odds with the positivist normative concep-tion of health deriving from Comte. In so doing, Canguilhem gestured toward new notions of singularity and normality. For him, the healthy needed to be rethought as "more than normal": that is, health is character-ized by the absence of symptomatic sensations, but is also an exuberance that is not limited to and by norms, but indeed constructs them. Disease then is a situation in which human beings are tethered to norms, often new norms, but nonetheless ones that affect suffering, the efforts to escape from it, and the now-glorified memory of health; disease makes it impossible to live without constant reference to norms and to deficiency and failure vis-à-vis these norms.[10] Canguilhem charged that the still-dominant positivist medicine, and positivist conceptions of disease and biology, misunderstood this, erasing individual reactions to disease, considering them merely as aberrations from normality that could or could not be corrected; in so doing, they also effaced the experience of suffering and even of health itself.

This argument resonated. Canguilhem's claim that a normativity based on statistically engendered normality fails to explain the complexity of phys-iological disorders and the individual patient's relation to his or her envi-ronment connected with the work of contemporaries—from Walter B. Cannon to the endocrinologist Hans Selye—who had begun to investigate the physiological grounds of what had classically been considered psycho-logical reactions (fear and anxiety, for Cannon; stress, for Selye; and so on). As scholars have recently shown, other physiologists (Karl Rothschuh) and medical thinkers (Pedro Laín Entralgo) should be included in this list; the same could be said for psychiatrists, from psychoanalysts concerned with the question of healing, to psychiatrists (Henri Ey) and even antipsychia-trists (such as R. D. Laing), whom Canguilhem would come to criticize.[11] It could even be said that *The Normal and the Pathological* belonged to a moment of intellectual recalibration in medicine when questions of pathological experience and of the complexity of biological conditions were becoming increasingly clear.[12]

In the context of epistemology and French philosophy, *The Normal and the Pathological* also echoed the work of a new generation of philosophers of science (Alexandre Koyré, Jean Cavaillès, Gaston Bachelard) who had questioned the status traditionally accorded to experimentation and the fantasy of positivism that they saw as defining much contemporary science and scientific thought. Canguilhem knew Koyré's problematization of the history of science well and found considerable inspiration in it for his own work.[13] He also shared much of his intellectual upbringing with Cavaillès, as well as a life-defining experience in their resistance to German occupation (Cavaillès would be executed in 1943); after the *libération* he became Cavaillès's chief promoter, editing and publishing his *Logique et théorie de la science* (*Logic and the Theory of Science*, 1947).[14] Canguilhem would also write on both Cavaillès and Bachelard—whom he replaced at the Sorbonne.[15] In many ways, the lineage to which Foucault would famously appeal in his link of Canguilhem, Bachelard, Cavaillès, and Koyré was elaborately constructed by Canguilhem himself.[16] Today, moreover, Canguilhem's epistemological and historical writing is read together with those of other important contemporaries, notably Ludwik Fleck and Thomas Kuhn.[17]

Canguilhem's book could be further read in a third context of intellectual-historical significance—the retheorization of the role and force of norms and normality that is an essential aspect of several twentieth-century schools of French thought, from Durkheimian sociology (for example, in Emile Durkheim's *The Rules of the Sociological Method* and Maurice Halbwachs's thesis *Théorie de l'homme moyen* [*The Theory of the Average Man*])[18] through psychoanalysis (significantly in the work of Jacques Lacan), the Annales school of historiography, and, of course, the writings of Foucault.

Canguilhem's second book and minor doctoral thesis,[19] *Knowledge of Life*, expanded his focus to include the history of biology and the interplay of philosophy, modern science, and conceptual history. Utilizing a perspective influenced by Henri Bergson's *Creative Evolution*, Canguilhem expounded on the irreducibility of "knowledge" and "life" to one another, addressing the political implications of cellular theory, the imperiousness and insufficiency of the mechanist legacy, and the intellectual history of the relationship between an organism and its milieu. Writing in the heyday of behaviorism, at the dawn of cybernetics, and on the eve of the genetic revolution—movements that would offer new paradigms for biological thought—Canguilhem used his conceptual histories to argue for a kind of *negative vitalism*: while rejecting vitalist science (both contemporary Lysenkoism and varieties of *Lebensphilosophie* from 1890 through World War II), he refused any reduction of life to mechanistic physico-chemical principles, instead seeing vitalism as a continuing impulse that demonstrates the profound violence that mechanistic biology does to the individual organism. Canguilhem argued that mechanism, dating back to Descartes, fails to see that particular milieus are always experienced differently by particular organisms and that organisms, because of their unpredictable and singular interactions with their milieus, cannot be reduced to undifferentiated physico-mechanical systems.[20] From this starting point, he developed an understanding not of "life"—a concept too generic and crude to be productive—but of "the living," the "living being," or the "living organism," phrased in a way that silences the possible organismic or ontological focus of this term "life." "The living," *le vivant*, differs further from "life," *la vie*, in that it is mediated, *as* living, by the specific milieu it experiences. Nonliving beings have no such milieu, whereas living ones are forced to deal with theirs—and their milieus are always different, specific to themselves. As a concept, "the living" allowed a rethinking of individuality in terms of both the indivisibility and the inherent differences between different beings, with normativity seen simply as a value imposed by living. At the same time, it avoided the vitalist implications of a "vital spirit" or a Bergsonian *élan vital* ("vital impulse").[21] This critique of mechanism would become a premise and focal point of Canguilhem's major doctoral thesis in *La formation du concept de reflèxe* (*The Formation of the Concept of Reflex*) and his later writings on behaviorism. Mechanism and the concept of reflex would serve as exemplars of what Canguilhem came to refer to as "scientific ideology."[22]

The idea of scientific ideology contributed to the first peak of Canguilhem's influence, in the 1960s, which related in large part to various phenomenological and Marxist debates on science. As has been amply documented, certain strands of French phenomenology took a scientific turn in the 1950s, thanks not only to Merleau-Ponty's persistent engagement with Gestalt theory and quantum mechanics in his 1956–57 lecture course, "Nature,"[23] and elsewhere, but also to the reception of Husserl's *The Crisis of European Sciences*[24] in the writings of Suzanne Bachelard, Paul

Ricoeur, Jacques Derrida, Jean-Toussaint Desanti, and others. In this context, Canguilhem's claim that scientific notions and practices are constructed thanks to an empirical and conceptual back-and-forth, but coalesce around theoretical systems that tend to become reductive and rigid, as well as his showing how the sciences define norms and frame experimental inquiry, echoed the Husserlian effort to handle ideal objects that have an empirical origin.[25]

In this period, Althusser and Althusserians—but also Bourdieu and others within his school of sociology—were attracted to Canguilhem's writing above all both because of its rejection of positivism and because of its distance from Hegelian concepts of truth and knowledge. Canguilhem countered the positivist and Hegelian stances by theorizing the sciences as systems that tend to be self-enclosed and tend to produce norms of their own, in order to formulate, validate, and defend what they perceive as their own "truths." He understood such practices as ideological and criticized them for ignoring the anchoring of scientific endeavor in social context and conceptual history. Offering a history of cell theory in *Knowledge of Life*, for example, he sought to show a sort of feedback loop between social theories and the various stages of experimental work on the cell as the indivisible living unit. The point was that in its search for the cell, biology at its various stages since the eighteenth century had played out various theories of individuality and sociality—for example, highlighting the "autonomy" of parts from the whole in Buffon's proto-cell theory and an organism's communal priority and superiority over the individual units' "realities" in Romantic biology (for example, in the work of Lorenz Oken and Matthias Jakob Schleiden).[26] Cell theory in its development from the work of Rudolf Virchow onward produced arguments concerning individuality that it deployed across a broad conceptual matrix far exceeding its particular experimental interests, from life, to autonomy, to subjectivity, to ontogenesis, all of which bore social and political implications and reflected social and political milieus.[27] Turned into a foundation of biology in general, cell theory, considered as a scientific ideology, thus gradually enforced a hegemony over both research possibilities and the social and political implications of scientific research. To practice the history of science, Canguilhem argued, is not to offer a narrative of bygone eras of science now known for their errors. Rather, it is to destabilize a *savoir*, a system of knowledge that has proclaimed itself to be the only fount of truth and that has had the power to label its scientific and political opponents' "errors" as disingenuous dwellings in fallacy and delusion.[28]

A skeptic worried by biological and physical determinism; a historian engaged with the question of the value of the living and invested in defending it against rationalization; a philosopher for whom individuality signifies irreducible novelty, a particular, semi-autonomous relation to the milieu, and a difference from other beings that would be structurally the same: across his writings, Canguilhem elaborated a philosophical modernism based on medical and biological individuality. By philosophical modernism we mean that already in *The Normal and the Pathological* Canguilhem found himself at a specific crossroads of modern thought, which he addressed by moving partly against classical subjectivism—which in his eyes has depended on a normativity radiating from the normal individual—and at the same time toward a radical, even an-archic, antifoundationalist concept of individuality and subjectivity.[29] By the time of *Knowledge of Life* and *The Formation of the Concept of Reflex*, Canguilhem was explicitly criticizing technological and scientific modernity for being conceptualized and structured in terms that ignore and destructively reduce differences between individual living beings and that thereby threaten their humanity.[30] This characteristically modernist stance relates closely to those of very different philosophers—from the late Husserl of *The Crisis of the European Sciences*, to the Horkheimer and Adorno of *Dialectic of Enlightenment*, to the Arendt of "Ideology and Terror" (the second-edition epilogue to *The Origins of Totalitarianism*).[31] However improbable this list may seem, a basic shared argument predominates: modernity has elaborated scientific and technological systems that eradicate man's wonder at the world, reduce truth to their rules and forms, and facilitate an at once political and ontological framing (if not quashing) of the particular, threatening a conception of the human that such philosophers wished to rearticulate and defend. In other words, work in the philosophy of medicine was for Canguilhem a means to further a political thinking of subjective existence *amidst* medical, technological, administrative, and political milieus. This metapolitical engagement defined Canguilhem as an intellectual, and not as merely a philosopher and scholar. It would be up to the later Canguilhem—indeed, the Canguilhem of the essays included in *Writings on Medicine*—to establish the more precise contours of his positive argument about the relation of the individual and social-medical systems.

It is nevertheless imperative to note at this point that for Canguilhem, knowledge and the existence of norms are decidedly *not* a problem: norms are essential to human life, and knowledge is an integral part of the human experience of the world. Science and scientific ideologies are ways in which this life becomes comprehensible to us. The problem that emerges is that norms and knowledge, tools thanks to which we experience and understand the world, are also tools that reduce this world and human life in general. In other words, understanding the limitations, reductiveness, and indeed danger inherent in norms and science is essential to understanding their force and creativity.

II.

Writings on Medicine is a collection of posthumously republished essays. The French edition of the book was included in the collection Champ Freudien at Éditions du Seuil and was organized by Armand Zaloszyc, a psychiatrist and psychoanalyst who had studied in part under Canguilhem. Because the reader might at first glance consider the collection arbitrary (there are many other essays by Canguilhem that remain spread out among their original places of publication and have not been anthologized), it is important to note that Canguilhem himself had drawn up a very similar publication project in the 1970s.[32] His outline imagined a book that would be titled *La maladie et le malade, la médecine et le médecin* (*Disease and the Patient, Medicine and the Doctor*) and that would include many of the same essays and lectures.

Canguilhem's outline includes "The Idea of Nature"[33] and "Is a Pedagogy of Healing Possible?"[34] which appear in the same form as Chapters 1 and 4 in the present collection. The book would further use his text "Corps et santé," a lecture for philosophy DEA students (the equivalent of a master's degree) at the University of Paris–I dating to April 29, 1977, which Canguilhem shortened a bit for the 1988 lecture "Health: Popular Concept and Philosophical Question,"[35] which serves as Chapter 3 here. "Corps et santé" also includes elements and formulations that Canguilhem elaborated in Chapter 2 here, "Diseases" (1989).[36] Two additional essays described in Canguilhem's book outline do not appear in *Writings on Medicine*. These are "Puissance et limites de la rationalité en médecine" ("The Power and Limits of Rationality in Medicine"), which was eventually included in the fifth edition of his *Études de l'histoire et de la philosophie des sciences* and shares rather little in terms of style and practice with the essays found here, and "Pour les dentistes" ("For Dentists"), a short 1979 lecture that was largely recuperated between the different essays here.[37] One essay Canguilhem did not consider including is Chapter 5 of the present edition, "The Problem of Regulation in the Organism and in Society."[38] This essay antedates the others and is somewhat different in style; moreover, as we will discuss later, Canguilhem later revised somewhat his thinking on this topic. Nevertheless, the essay's main themes echo the concerns and to a considerable degree the style of the overall volume, and as we will see, were of major concern for Canguilhem.

We do not point to Canguilhem's project description in order to suggest that the idea for a book like the present one was his, but to insist on the fact that the essays included in the present collection follow a precise rationale. The rationale involves a number of axes that frame Canguilhem's philosophy of medicine and his specific way of using history. Though this rationale has much in common with Canguilhem's other works from the 1970s, it is only through the essays that form *Writings on Medicine* that it becomes consistently and explicitly elaborated. Here we find a philosophy based on an elaboration of medicine's fundamental bases for therapeutics that uses often taken-for-granted aspects of medical and scientific thought. As we will see, Canguilhem deals with highly particular topics in his essays, yet in doing so attempts neither to establish a new ground for therapeutics nor simply to recuperate traditions of healing, such as those following Hippocrates, Galen, or Xavier Bichat. His accounting aims instead to create an understanding of therapeutics based on philosophical registers that already exist, but that, for whatever reason, remain tacit. It is here that the rationale behind his broader philosophy of medicine becomes clear.

Four major medical-historical axes structure Canguilhem's argument. First, in the wake of increasing technicalization and professionalization, Canguilhem argues that it is essential to rethink medicine, beginning with its classical determination as a set of techniques and practices aimed at curing the ailing body. In this way, medicine also needs be considered to be a technology of healing individual bodies, not merely a system or applied science of *the* body.[39]

Second, the "naturalism" accorded to the human organism must incorporate the ways that place, circumstance, and society come to bear on this idea. The body is only quasi-natural; it is profoundly responsive to and affected by the conditions of human life, situated in the context of a society. As a result, and given modern technological change, appeals to nature and to a Hippocratic tradition aiming to restore the body to its natural health are as philosophically retrograde as they may be medically counterproductive. When medicine fails to "cure," this does not mean that nature will step in instead.

Third, the values attached to "health" and "disease" must be challenged and rethought, particularly when these values form a moral register of "good" and "evil." To start with, the pursuit of health may require conniving with disease—and as a result, the traditional valuation no longer suffices.[40] More importantly, and precisely because health and disease involve a profound valuing of the individual—by the individual, as well as by society—articulating these values anew and doing so in a philosophically and historically careful way has become imperative. Health as a value must be inflected by finitude and time: first, because health cannot be seen as permanent, and second, because it is a factor of temporal process, because the body declines and never quite "returns" to an anterior living state following a disease. This lived sense of health and disease is crucial for the patient, who finds his or her expectations and hopes managed by medicine, just as it is for the physician, whose efforts to cure are necessarily constrained by the patient.

Fourth, the institutionalization of medicine has led to a cheap abstraction of "the subject" and woeful generalizations of "the individual." In regard to this last point, Canguilhem is exacting when he argues that the tension that arises between "individual" and "collective" medicine must be examined and perhaps challenged, especially when it begins to threaten the individuation of experience. An organism is a priori and necessarily a negotiation of individuality. In developing externally confirmed systems of cures, medicine tends to take this individuality for granted, when it is not guaranteed. Nor is it a shibboleth: emphasizing individuality should not lead to a medical subjectivism or to a trust in the body's self-regulation. Individuality is precarious, crucial for the patient who understands himself or herself as an individual and not as a subject for medical research and practice, and difficult to handle from the perspective of the doctor-patient relation.

It is through these lines of thought that Canguilhem creates a coherent link between the history of medical thought and a philosophy of medicine—a philosophy of therapeutics, a philosophy of an organism threatened by its milieu and uncertain of its ability to handle disease and medical intervention.[41]

The specificity of *Writings on Medicine* resides also in the way in which Canguilhem "does" history. For him, medical thought, medical practice, and the experience of health are historical objects. In history, we discover their traditions, aims, and efforts, but also the conceptual frames within which they present and bear out philosophies of medicine. At the same time, historical practice serves as the process for philosophical engagement and contemporary political engagement. And though Canguilhem draws upon traditional medical historians to trace the movements of ideas and practices, here he is not attempting to reconcile contemporary biomedicine with its history.[42]

Canguilhem's practice in *Writings on Medicine* is a mixture of what we usually refer to by the terms "conceptual history," "historical epistemology," and the "history of the present." Once again, while this mixture is not unique to *Writings on Medicine*, it is carried out with particular insistence in this book.

Conceptual history should not be understood here in quite the sense given it by Reinhard Koselleck's practice of a *Begriffsgeschichte*, yet the term is nevertheless essential as a qualification of Canguilhem's engagement with the history of ideas. What is at stake is not so much the history of ideas, considered in the "Platonist" sense as dissociated from society and culture and having motivations and implications only of their own.[43] In this approach, concepts put into practice by medical thinkers, philosophers, and other cultural agents participate in broader conceptual and notional frameworks. A history of concepts viewed in this way doubles as the history of their definitions, their related figures, the matrices to which they belong, and the concepts with which they bind, including others with which they have been contrasted or from which they have been contradistinguished. Since *The Normal and the Pathological*, this engagement with concepts and the practices that surround them has been central to Canguilhem's thought. In the present collection, and indeed more so here than in some of his other works, Canguilhem extricates concepts ("nature," "health," "disease," "pedagogy," "cure," "regulation") from the matrix that would link them directly to one another in order to treat the way in which each bears a singular history—a history that renders their binding together less stable than previously assumed.

Historical epistemology can be defined as a history of the production of knowledge or as a line of questioning that considers the history of concepts and often-opposed scientific approaches as developing systems of knowledge and of the production of truth.[44] Hans-Jörg Rheinberger has described Canguilhem's practice as forming "a history of the displacement of problems which must be reconstructed in their historical context."[45] In other words, central to his thought and historical practice is the question of how historical agents come to know what they know, how they come to articulate it, how they play out presuppositions of their thought in their practice and scientific writing. Put differently, central to his thought and historical practice is the question of how particular problems (disease, cure, and health, above all, in the present case) have a lived as well as conceptualized significance. As he writes in "The Object of the History of Science:" "The history of science concerns an axiological activity, the search for truth. It is at the level of questions, of methods, of concepts, that scientific activity appears as such."[46] Canguilhem's specific engagement at the level of questions, methods, and concepts takes the form not of tracing epistemological breaks in medical thought (what Bachelard described as "ruptures" or "coupures épistémologiques"), but of a series of historical continuities that compete with one another in articulating the terms in which the above-discussed questions that frame Canguilhem's philosophy of medicine can be asked. This book as a whole brings a series of conceptual histories and the practices they have facilitated to a contrast that elucidates and dislocates their implicit force within scientific systems.

History of the present is a term Michel Foucault proposed in his 1975 *Discipline and Punish* as his way of thinking about the link between the nineteenth-century focus of his "genealogical" studies and the contemporary intervention he sought to make through them. Canguilhem's essays in *Writings on Medicine* make a distinct effort to intervene in contemporary debates in medicine, sometimes through short asides citing policy proposals, but more frequently through a constant sense that health, disease, cure, nature, and individuality are fundamentally problems whose history remains implicit and troubling in their contemporary uses. Canguilhem's practice, in other words, is motivated by the standing of the historian of science as a historical agent in his own right, and thus one whose truths and axiological investments are reflected in his practice, as well as by the urgency of a historical moment in which he feels a lack in the historical understanding of medical concepts to be inflecting, if not deforming, the practice and goals of medicine. Interpreting Canguilhem's own practice as a history of the present thus also brings to the fore the political dimension of the rationale that pervades the essays included here—political at least in the sense that it concerns the place and politics of medicine in a world increasingly dominated by a far-fetched rationalism.

Nowhere is this clearer than in Canguilhem's harsh critiques of antipsychiatry and antimedicine, movements whose claims were widely felt in the 1970s. Crucially, Canguilhem does not elaborate a critique of systematic medicine in order to refuse medical practice and conform to a notion of the medicalization of life that we find in the work of Ivan Illich, the infamous author of *Medical Nemesis* and a scholar whose antimedicine Canguilhem criticizes harshly and rejects.[47] Indeed, nothing could be further from the argument here than that, given the repressiveness and disrespect of subjectivity to be found in modern medicine, one should instead opt for a supposedly natural and unmediated, unrepressive notion of health. Canguilhem is not hostile to medical practice and experimental science—his claim for a history of the present lies precisely in the critique of the frequent ignorance of basic presuppositions in medical thought that allowed antimedicine to come to the fore.[48] Thus, what sparks Canguilhem's critique of medical reason, what keeps these essays together as a unit, is not a wish to declare medical systems insufficient and to aim for a natural or Hippocratic idea of health. Quite to the contrary: it is a demand for ways of reinvesting in modern medicine, with all of its limitations, and explaining in what way it holds therapeutic force.

III.

On the basis of his philosophy of medicine and his historical practice, Canguilhem identifies each chapter with one or more overarching questions. How are we to contend with the "naturalism" that is imagined in medical thought and practice (Chapter 1)? What is the value of disease in relation to healing and to curing (Chapter 2)? If "health" were understood as an a priori concept, then where would we find its foundation (Chapter 3)? Are there notions of healing that can be held apart from the concept of a "cure"? And along similar lines, what are the limits of psychology in medical thought?[49] Can we find a register for the experience of illness and suffering within current conceptualizations of human biology and physiology in medical practice (Chapter 4)? And finally, what is the relationship between the organism and society and how do theories of the organism's regulation renew the philosopher's engagement with this question (Chapter 5)?

Chapter 1, "The Idea of Nature," focuses on the Hippocratic tradition of understanding nature as the degree zero of health. Canguilhem foregrounds the "healing power of nature" (*vis medicatrix naturae*) in its presumed analogy to and superiority over the practice of doctors, and he traces three traditions. The first is the tradition of Western medicine's suspicion of nature's healing power. The second is the motif of a "medicine without a doctor" in the seventeenth and eighteenth centuries, the idea that "natural man" is "his own doctor"—a recuperation of the Hippocratic tradition that Canguilhem sees at the base of the fascination with a "natural health" in modernity. The third is contemporary medicine's complicated engagement with the Hippocratic tradition, which has resulted from the discovery that we often need to use nature against itself if we are to (re)gain health. Canguilhem's stance toward "nature" is ambivalent here—largely because of his mistrust of then-growing appeals to a vague "nature." In his essay "Nature naturante et nature dénaturée" ("Naturing Nature and Denatured Nature"), which appeared four years after "The Idea of Nature," in 1976, Canguilhem lamented the difficult position that a philosopher who would reject the naturalisms then "à la mode" found himself, and he proceeded to adopt just that difficult position, denying the simple opposition imposed by the expression "denaturing."[50] Furthering the title's allusion to Spinoza's concept of *natura naturans*, that is, of nature generating itself, Canguilhem argued that nature can also *de*nature, that it can denature itself, lose the elements that have come to define it as such.

> Because nature can only be naturing, a denatured nature—at once a daughter and a mother to culture—is possible. And because this denaturation progressively had to borrow the paths of abstraction and of nonfigurative representation that belong to science, and as a result one could not recognize naturing nature in a denatured nature, a complaint and an anger were born to which literature and ideology vainly seek to give philosophical weight.[51]

In "The Idea of Nature," Canguilhem similarly articulates the demand for a philosophical and historical understanding of the concept of nature within the different traditions that he engages. Naturalism offers nothing by itself—merely a mistaken rejection of medicine and through it of the forms of care that engage in a complex fashion with nature. The broader philosophical implication here is that nature cannot acquire foundational status in philosophy. Canguilhem denies that human beings are somehow fundamentally sustained by nature alone, or that trust in such sustenance can, today, suffice as anything more than a basic critique. A final aim of this argument is to open up the space for a judgment on the doctor-patient and doctor-nature relations with which Canguilhem will engage in the later chapters.

In Chapter 2, "Diseases," Canguilhem begins with the Hippocratic tradition to identify "the instructive work" of illness in order to situate disease as a purposeful, "reasoned" activity of the body, one that, at times and through its evidence in symptoms, invites intervention. Rather than simply tracing the chronological progression of an understanding and valuing of disease throughout the course of medical history, in the chapter Canguilhem elaborates on its status as a part of medical practice (aiming to manage symptoms and to rid the body of the cause of disorder) that has come to organize and fix a so-called "medical perspective." Canguilhem draws attention to the conceptual distances that separate "disease" and the "sick," what he calls the "patient-disease" (*malade-maladie*) relationship, a relationship that, in his view, cannot be of complete discordance. In Canguilhem's formulation, being sick is a norm of life, and disease only establishes a new biological norm "at the nebulous border between somatic medicine and psychosomatic medicine," resulting in "both deprivation and change."[52] Canguilhem engages the concept of disease—its value and role—less in an attempt to recover a lost holism than as a beginning of what could be called a "philosophical anthropology" of medicine, one that puts analytical historical pressure on the self-evidence of medical thought.[53]

In many ways, Chapter 3, "Health: Popular Concept and Philosophical Question," extends the line of questioning laid out in Chapter 2, "Diseases." As mentioned earlier, both of these chapters (and "Health" especially) derive from the lecture text "Corps et santé" ("Body and Health"). They also extend arguments that Canguilhem had begun when considering the teachings of the physician René Leriche in the first version of *The Normal and the Pathological* (1943) and then sharpened in the chapter "Disease, Cure, Health" of that book's extended second edition (1966). Chapter 3 is driven by a single polemic: if "health" exists with reference to some explicit knowledge, where do we find its ground? Canguilhem explains that "health" has remained a topic of philosophical thought throughout the works of Kant, Nietzsche, and Maurice Merleau-Ponty. However, it is Leriche's formulation regarding health as "life lived in the silence of the organs" that Canguilhem places under the most careful scrutiny.

Here we begin to see the nature of Canguilhem's relationship to Friedrich Nietzsche—Canguilhem called himself a "nietzschéen sans carte," a non-card-carrying Nietzschean.[54] Nietzsche, who also found inspiration in Claude Bernard, did not accept "health" as unproblematic. For Nietzsche, the concept of health—"the great health" (*die große Gesundheit*)—is at once corrupted and corrupting; it misattributes virtue; it paradoxically serves as a ground for malice and a conditional life.[55] It is in this chapter that we also find a reaffirmation of the importance of Kurt Goldstein's conception of health, which postulates the positive interlocution between a living individual and its milieu, a milieu that is often *ébranlé*—threatening—and not secure. For Canguilhem, "health is a margin of tolerance for the inconstancies of the environment," and the "healthy" organism is one that follows a "privileged" path through its environment and does not behave in a disordered or (worse) "catastrophic comportment."[56] Canguilhem further links Goldstein's treatment of catastrophe to Walter Cannon's (1914) writing on the problem of "alarm,"[57] Hans Selye's 1936 work on the general adaptation syndrome to nonspecific stimuli, a theory of "stress,"[58] as well as René Thom's theory of catastrophe.[59] As Canguilhem had already recounted, Goldstein's argument involved precisely a definition of health as the ability to not be caught in a relation of *ébranlement* with the environment, but to be able to overcome its dangers.[60] Yet Canguilhem shifts focus to consider not only the physical world, but the social one. In order to define health, he attends to the alignment between the organism and its milieu, both physical and social. Managing the social aspect of health (as he recounts with reference to tuberculosis) is just as crucial as managing its physiological aspect. It should be clear by now, moreover, that in engaging with health, Canguilhem finds a central way of approaching subjectivity, and the place and status of the subject within nature and society. Having

refused both the sociological and pure philosophical approaches and having taken from but elided both existentialism and structuralism, Canguilhem treats health and disease as privileged entry points for a thinking of
the subject and self within social, psychological, and biological frameworks, a subject either exuberant toward itself and its world (healthy)
or constrained, bound to norms it does not control or adequately handle
and potentially in a direct relationship with decline and imminent death
(diseased).

Chapter 4, "Is a Pedagogy of Healing Possible?" utilizes some of the
central insights of the preceding chapters—that an organism's natural state
can be harmful to it given its current circumstances, that diseases are instruments of life, but are seen to be so only with difficulty given the harm they
inflict on the organism, that a cure is not a restoration of an anterior
state—to address the role and function of the physician. Demonstrating a
fundamental difference between externally confirmed cures and subjective
healing, Canguilhem asks how healing and curing can be theorized while
keeping in mind the perspective of the patient, who refuses simply to accept,
as his or her own, the interpretation of the cure offered by the doctor. At
stake, then, is medicine's role in subjectivity itself.

The essay is anticipated in part in a lecture on therapeutics that Canguilhem had delivered in 1959, "Thérapeutique, expérimentation, responsabilité" ("Therapeutics, Experimentation, Responsibility"), in which he
articulated in some detail the divide between, on the one hand, experimental
and technologically grounded medicine, and, on the other, the needs of therapy, understood as the restoration of a healthy state to an organism that has
been damaged or endangered.[61] There, Canguilhem had largely engaged
with Immanuel Kant's *The Conflict of the Faculties* in order to articulate a basic
claim that philosophy's meddling in medical matters can only go so far in its
judgments, while medicine needs to be "radically desacralized."[62] He
accepted the authority of experienced and cultured physicians and called for
the establishment of a superior way to initiate the teaching of medicine—in
a sense setting up a part of the argument he would pursue here.

Yet at the same time, the problem of "Is a Pedagogy of Healing Possible?" is rather different and could perhaps be best understood as an implicit
transposition onto the doctor-patient relationship of the psychoanalytic
problem of transference. Therapeutics, Canguilhem argues, requires something that is often impossible—the meeting of a medical cure with social as

well as individual healing. If the doctor's success is not only a matter of "fixing" the organism, mechanically, but of restoring its relationship to its milieu, then situations of serious illness all but prevent such a restoration. Given that much, the doctor-patient relationship is not an a priori failure, but it is inflected by a mixture of conditional, partial therapeutic success with a recognition of the finitude of individual goals in a given environment and the transformation (both individual and social) that the patient will necessarily endure. And such a transformation comes to require a holistic approach to patienthood.

A basis for therapeutics, in this argument, comes closer to the terms to be found in the writings of the renegade psychoanalyst Georg Groddeck[63] and the then-surprising work of the psychoanalyst and homeopathic physician René Allendy. Best remembered for having treated Anaïs Nin (and Antonin Artaud), Allendy was an early influence on Canguilhem through his insistence that healing cannot be separated from a concept of the medical subject.[64] For Canguilhem, Groddeck and Allendy recognize the need for a cure to become, in medical eyes, "the sign of the patient's rediscovered capacity to be finally done with his own difficulties." Like Nietzsche, they point to a conceptual transformation that Canguilhem deems necessary so that human biology and society can become more effectively bound up with the performance of healing, so that scientifically based care does not fall prey to appeals to personal healing—to "the first therapist to come along and appeal to psychosomatics." They also resonate with Canguilhem's continuing attention to Goldstein, who provides not only the crucial spark for an analogy between therapeutics and pedagogy, but also the persistent commitment, far from any easy medical humanism, to a doctor-patient relationship defined as "a coming to terms of two persons, in which the one wants to help the other gain a pattern that corresponds, as much as possible, to his nature."[65] It is thanks to this commitment that Canguilhem can call for a Critique of Practical Medical Reason, whose purpose would be to balance, compare, and weave together experimental and scientific knowledge with the "propulsive nonknowledge" of a living being's effort to survive and retrieve a life in health—a life and health that neither knowledge nor life can institute or guarantee.

Chapter 5, "The Problem of Regulation in the Organism and in Society," dates to 1955, some twenty years before the other essays and at a time when Canguilhem's intellectual priorities differed considerably from those he would adopt in the 1970s. It can perhaps be seen as an easier and more explicitly politicized effort than Canguilhem attempts in the other essays included here. What helps in understanding his approach, however, is not only that it was directed at a popular audience (at the Alliance Israélite Universelle), but that at the time, Canguilhem was elaborately reworking his stance on the problem of how the organism is regulated and that, notably, his understanding of regulation would come to play a major role in his understanding of the individual organism and, through it, of health. The lecture thus offers an early glimpse at Canguilhem's effort to grapple with the idea of regulation, and it is worth attending at some length to the ways in which his thought on the matter would be finessed over time.

In "The Problem of Regulation in the Organism and in Society," Canguilhem recuperates the importance of an organism's self-regulation as a determining element in its constitution. With its examination of Claude Bernard's notion of the internal milieu and of Walter Cannon's understanding of homeostasis, the essay clearly expands on "The Living and Its Milieu" in *Knowledge of Life*, writing that "the originality of Claude Bernard resided in his showing that . . . it is the organism that produces this internal milieu. I insist here on the fact that the regulation of the organism is ensured by the special devices that are the nervous and endocrine systems." The priority of the organism is crucial, because, as in "The Living and Its Milieu," it allows not only for Canguilhem's emphasis on the individuality of the organism, but also for a suggestion that such individualized regulation is at the core of the organism's self-construction. On this basis, Canguilhem proceeds to show the differences between organisms and societies, which lack the natural givenness of organismic regulatory mechanisms.

It is important, still, to note that, in the ensuing argument that points out the dangers of expanding regulation to a social principle, the organism's self-regulation becomes the foundation for health. Regulation, in other words, becomes a crucial element in Canguilhem's effort to overcome Cartesian mechanism and to offer a medically informed philosophical theory of the body. This is, indeed, what places this chapter at the base of *Writings on Medicine*. In assuming this major role, however, regulation remains shrouded in ambivalence—an ambivalence that would only be heightened, thanks to the rise of other theories using regulation in the postwar period (notably cybernetics) that would put further stress on Canguilhem's prioritization of regulation. In "The Problem of Regulation in the

Organism and in Society," Bergson serves as a counterpoint to the Bernard-Cannon line of development of the theory of regulation, occluding any strict commitment to such a theory. This use of Bergson is particularly significant, insofar as Canguilhem's criticism of Cannon is premised on both his influence on Arturo Rosenblueth and Norbert Weiner (whose cybernetics remains a quiet target of this paper) and his extrapolation from physiology toward a social theory. A commitment to regulation, Canguilhem suggests through Bergson, amounts once again to an elision of individuality and to an ordering that is philosophically dubious and might become politically destructive.

Canguilhem would broach the subject of regulation again in 1957–58 in his weekly seminar on the history of biology at the Institut de l'Histoire des Sciences.[66] In a way commensurate with "The Problem of Regulation in the Organism and in Society," his seminar argument on regulation addressed the relationship between an organism and its milieu as this developed in the nineteenth century (notably in the work of Comte) and the shift from external to internal milieu that occurs with the work of Claude Bernard, also as suggested in "The Problem of Regulation in the Organism and in Society." Bernard's notion of the internal milieu provided at once a sense of the organism's unified response to stimuli and a sense in which this response is both guided from within and responsive to internal as well as external disruptions. In the seminar, Canguilhem further insisted on the importance of Cannon's studies of "the importance of the autonomic nervous system on the regulation of physiological functions" as a form of internal self-regulation and revisited much the same set of historical origins as he does here: the Hippocratic *vis medicatrix naturae*, for example, but also the biologists Eduard Pflüger, Leon Fredericq (on animal heat), and Charles Richet. He also addressed, for the first time, the contribution of cybernetics to the return of the term *regulation* and to its role in the imagination of a self-regulated society, pointing (without recorded commentary) to W. Ross Ashby's 1948 Homeostat, his device intended to simulate the brain's regulation of disturbances. Here again, Canguilhem's suspicion of self-regulation—or mechanical organization, for that matter—remains core to his sense that human life—and perhaps life itself—cannot be reduced to an order without unexpected and uncontrolled disturbance.

Canguilhem then returned to the concept in the 1970s. In 1971, he published an article on regulation in the French *Encyclopaedia universalis*[67] that again emphasized the lineage of Bernard and Walter Cannon, but also offered a broader, tripartite account of the epistemological, biological, and sociological use of the term. Soon thereafter, in 1973, he presented a strongly political version of the same article in a three-lecture course at the Catholic University of Louvain titled "Fin des normes ou crise des régulations?" ("The End of Norms or a Crisis of Regulation?"), which he concluded with the lecture "Regulation as a Reality and as a Fiction." Preoccupied with the emergence of antipsychiatry and antipedagogical movements toward the end of the 1960s, Canguilhem mounted an attack on both normative and antinormative thought—that is to say, not only against normalizing tendencies in society but also against the fantasy of overcoming hierarchies (such as doctor-patient hierarchies) and rehabilitating ideals of "natural" health. Regulation as a mechanistic and cybernetic notion served to reimport into sociology and conceptions of the brain a mechanistic determinism and thus became a first target.[68] At the same time, we find Canguilhem extending his critique of antimedicine, which we have seen in Chapters 1 and 4: while maintaining a philosophical argument for the tradition that counters the medical establishment as source of sole medical truth, he distanced himself from hopes of "natural" [*sauvage*] medicine, and natural self-regulation in society.[69] Throughout, and once again, Canguilhem flatly refused Cannon's postulate in the concluding chapter of *The Wisdom of the Body* that an analogy could be derived from biological to social homeostasis, calling social self-regulation a myth and an ideology.[70] Against regulation, normativity remained, instead, the ineffable and not a priori undesirable basis of political and social reality.

Finally, Canguilhem would compose "The Formation of the Biological Concept of Regulation in the Eighteenth and Nineteenth Centuries" as a lecture he would deliver at the opening session of a 1974 conference at the Collège de France that engaged concepts of regulation in systems theory, biological and physical self-organization, genetics, and linguistics.[71] That essay would become a keystone chapter of *Ideology and Rationality in the History of the Life Sciences*, and it would return once again to the use of regulation in cybernetics and to the nineteenth-century elaboration of the concept of regulation itself.[72]

The evolution of regulation into a solid and potentially reductive theory of the organism thus ended up posing even at the level of individuality the very problems from which Canguilhem had used it to escape. Without leading him to give up on Bernard's concept of an internal milieu or Cannon's concept of homeostasis, the concept of regulation would remain, philosophically speaking, both promising and profoundly troubling. And in this, it is exemplary of the promise and insufficiency of all the concepts engaged in the current volume—concepts whose epistemological success breeds both an intellectual and a political excess of their epistemological value. In *Writings on Medicine*, medical reason remains bound to the often-contrary claims made for it by different philosophical-medical concepts and their histories. The conceptual equilibria it develops remain unstable, its solutions tenuous. Medical reason becomes by necessity its own critique.

IV.

In closing this introduction, a few notes on the translation are in order. Where Canguilhem translates from other languages, we have by and large used his renditions or amended the English editions. In keeping with recent usage (notably in translations of the work of Michel Foucault and Giorgio Agamben), we have translated *dispositif* as "apparatus" and have rendered *appareil* through the English terms "device" and "apparatus," depending on the particular case.

Vis medicatrix naturae: this Hippocratic expression is usually rendered in English as "the healing power of nature"; here, we keep it in the Latin as per Canguilhem's own usage, and we also keep in quotation marks "*la nature médicatrice*" when it appears; this signifies "healing nature" or "nature as medic," and we render it as "healing nature." Canguilhem also uses *médecine sauvage* to refer to naturalist medical treatments; though we might commonly refer to these by the term "alternative medicine" today, we have chosen to use "natural medicine" to avoid the anachronism and emphasize the link to various invocations of the *vis medicatrix naturae*.

Médecine expectante/expectation: The French term *expectant* has no simple English cognate. The quotidian use of the French adjective *expectant* would mean "attentive," "wait and see," or "passive," and the term is commonly used with reference to Hippocratic passive treatments. We have followed Svetolik P. Djordjević's *Dictionary of Medicine*, 2nd ed. (Rockville: Schreiber

Publishing, 2000), 524, in rendering *médecine expectante* as "expectant medicine"; at times we have also used "wait and see" as opposed to "passive," depending on the circumstances.

Maladie: Throughout the text we have translated *maladie* as either "illness" or "disease," depending on the context in which the word is being used. The distinction we have attempted to make is that in the case of "disease," the context is the discussion of a discrete object of medical inquiry or a specific entity or condition, while "illness," the context is the discussion of physical distress that is not bound by a specific condition or etiology of symptoms related to disease as such. It is worth noting that starting from Chapter 1, Canguilhem also plays on the polysemy of *le mal*—which points to "evil," "harm," "disease," and "wrong." No English term quite captures these meanings while working in context, so Canguilhem's uses of *le mal* will be highlighted in brackets.

Guérison: Our rendition of Canguilhem's term *guérison* as "healing" requires some explanation, particularly since Canguilhem makes an important and necessary distinction between "healing" and "cure," and we want to attend here momentarily to the complexity of each of these terms. "Healing" is fundamentally subjective and individual, following an etymology that includes protection and security, but also defense. Moreover, there is the possibility of confusion between older and more recent uses of the term. The older definition of the term *guerir* (after Alain Rey, *Dictionnaire de la langue française*) is *defendre, protéger*—"to defend, to protect." But the principal modern definition is *protéger, garantir*, now adding "to guarantee" and insisting that the proper sense is "to defend." To "guard against" is another example: the word is *guérite*, from *guarir* or *garir*, meaning *protéger*. The modern term contains the additional meaning of "protection" and "security" in reference to the political assertions of public health from the nineteenth and twentieth centuries in France. "Cure" (often also *remède*), on the other hand, reflects forms of internal change adhering to external validation. A "cure" here is the success of a change verified statistically or otherwise from without. However, the definition of "to cure" is also "to restore" (Latin: *curare*). Stated simply, as employed here, "cure" aims at a return, whereas "healing" opens onto something new and previously unfamiliar or unknown. We have attempted to keep Canguilhem's meaning intact. It is helpful also to remember the significance of the word "pedagogy" in Canguilhem's title "Is a Pedagogy of Healing Possible?"—"pedagogy" from the Greek, παιδαγωγέω (*paidagōgeō*), from παῖς (child) and ἄγω (to lead); literally, "to lead the child."

Finally, Canguilhem typically makes far more references than actual citations in his text. We have restored here these silent citations as translators' notes.

The Idea of Nature in Medical Theory and Practice

One may wonder whether the doctor-patient relationship has ever succeeded in being a simple, instrumental relation that could be described in such a way that the cause and the effect, the therapeutic gesture and its result, would be directly related one to the other, on the same plane and at the same level, without being mediated by something foreign to its space of intelligibility. It is certain, in any case, that the centuries-old invocation of a *healing nature* [une nature médicatrice] has been and remains the reference to just such a mediator, who would account, throughout history, for the fact that the doctor-patient couple has only seldom been a harmonious one, in which each partner can be said to be fully satisfied with the other's behavior.

To procure, for the sick man, by efficacious interventions, an amelioration or a restitution that he would not know how to obtain by his own means: that is the doctor's ambition, an ambition that is not ruled out by his sincere and persistent abstention from all charlatanism but is merely the other side of his professional honesty. This ambition may even contain the idea that a sick organism is, vis-à-vis the doctor and for him, nothing more that an object that is passive and obedient to external manipulations and solicitations. John Brown, a Scottish doctor much celebrated in Italy and in Germany at the beginning of the nineteenth century and the inventor of the concepts of stenia and asthenia,[1] believed it possible to summarize succinctly the imperative of medical activity: "In order to both prevent and cure diseases, we must always use the indication proposed, and stimulate or debilitate; never wait, nor trust the supposed powers of nature, which have no real existence."[2] This was the necessary consequence of a certain conception of the living body: "Life is not a natural, but a forced state . . . kept . . . not by any powers in themselves, but by foreign powers."[3] For an inert body, an active medicine.

Conversely, an awareness of the limits of the power of medicine accompanies any conception of the living body that attributes to it, in whatever form, a spontaneous capacity of conserving its structure and regulating its functions. If the organism has its own capacities of defense, then to trust it (at least temporarily) is a hypothetical imperative for both prudence and skill. For a dynamic body, an expectant, passive medicine [*médecine expectante*].[4] Patience—a wait-and-see attitude—would thus be the spirit [*génie*] of medicine. And still it would be necessary for the patient to agree to this forbearance. Théophile de Bordeu observed and wrote of this very well: "this expectant method has something cold or austere to it, and the vivacity of patients and assistants need not adapt to it. Also, those who employ it have always made up only a small number of the doctors, especially among people who are naturally sharp, impatient, and apprehensive."[5]

Not all treated patients heal. Some patients heal without a doctor. Hippocrates, who consigns such remarks to his treatise *The Art*, also holds the responsibility for—or the legendary glory of—having introduced the concept of nature into medical thought.[6] "Nature is the doctor of disease."[7] By "doctor," we should understand an activity, immanent to the organism, that compensates for deficiencies, reestablishes disruptions of the equilibrium, and corrects the bearing [*allure*] when it detects a change. This activity is not a thanks to a natural intelligence [*science infuse*]: "nature finds the ways and means by herself, not by way of intelligence: for example, blinking and the tongue offer assistance, as do other actions of this kind; without instruction or knowledge, nature does what is appropriate."[8]

The analogy between the art of the doctor and healing nature does not elucidate nature through art, but art through nature. The medical art must observe, must listen to nature. Here, to observe and listen is to obey. Galen, who attributed concepts one can call merely "Hippocratic" to Hippocrates himself, also took them up as his own: he, too, taught that nature is the first to conserve health because it is the first to form the organism. One must remember, however, that no Hippocratic text represents nature as infallible or all powerful. If the medical art has been born, if it has been transmitted, if it must be perfected, it is as a measure of the capacity of nature, as an evaluation of its forces. According to the result of this assessment, the doctor must let nature do its work, intervene to support and help it, or else give up on intervention—since there are diseases stronger than nature. Where nature gives in, medicine must give up. "For if a man demand from an art a power over what does not belong to the art, or from nature a power over what does not belong to nature, his ignorance is more allied to madness than to a lack of knowledge."[9]

Regrettably or not, the fact is that today, in order to practice medicine, no one is expected to have the least knowledge of its history. It is easy to imagine what impression a medical doctrine such as Hippocratism can make on the mind of someone who knows the name of Hippocrates only by the famous oath, the last surviving rite, which has now been emptied of its meaning. And it is worse still when, retroactively projecting into the past the theoretical principles and the technical precepts of current medical teaching, some take it upon themselves to judge Hippocrates as if what comes to us downstream in the course of history also must appear at the source. Let us note, without animosity, that even a master such as Édouard Rist, who was not at all ignorant of history, did not know in his *Histoire critique de la médecine dans l'Antiquité* (*Critical History of Medicine in Antiquity*, 1966) how else to treat Hippocratic medicine except by way of an indictment.[10] Apparently, this kind of ingratitude does not lack a foundation. As François Dagognet has shown in *La raison et les remèdes* (*Reason and Remedies*, 1964), contemporary medicine, far from systematically surveying or stimulating the organism's reactions of self-defense, often does its best to moderate, if not to suppress them, for example, when it tries to stop humoral reactions disproportionate to an aggression that causes them.[11] Sometimes therapeutics even collaborates with harm [*le mal*], reinforces what it should weaken, multiplies what it should reduce, in order to convert into an instrument of good the stimulation caused by a spontaneous reaction. Such is the case with some immunological practices that rely on the intensity of the infectious process to facilitate the action of bactericides by the secretion of proteolytic substances.[12] Doesn't it thus seem that contemporary medicine reverses the Hippocratic prescriptions and acknowledges the existence of a healing nature [*nature médicatrice*] only in order to fear and consequently to block its initiatives? This is because contemporary pathology has learned to see an orthodox Hippocratism in the existence of paradoxical organic reactions. Erroneous responses sometimes drive nature into a paroxysm over trifles. This is how things go in allergy and anaphylaxis. It is sometimes too little to say that nature's remedy is worse than the disease [*le mal*]—it is itself disease and harm, it is itself evil [*le mal lui-même*]. Yet if we examine closely the medical techniques that defend against this immoderate self-defense, doesn't it become possible to grant new meaning to the concept of nature?

Contemporary medicine practices an attitude of tentative doubt toward natural organic defenses. The doubt concerns not the fact of the reaction, but its initial pertinence and ultimate adequacy. And thus this doubt does not suspend the decision to intervene—on the contrary, it precipitates it. This is because this suspicion is founded on the knowledge of the role played by the autonomic nervous system in what have been called "pathogenic situations," independently of the nature of pathogens.[13] However, in the final analysis, any action on the autonomic nervous system, whatever its indirect mechanism or complexity of diversions—notably through the hierarchical inhibition of the centers of excitation or of braking—remains a copy (however reversed) of the natural organic process.[14] Even when it does so in reverse, art imitates nature in the sense that La Fontaine said that "My imitation is no slavery, I take only the idea, and the turns, and the laws."[15] A systematically non-Hippocratic therapeutics became possible around 1921, when Otto Loewi, confirming observations carried out by Thomas Elliot and Sir Henry Dale since 1904, succeeded in demonstrating that the vagus nerve acts by releasing an inhibiting substance, a chemical transmitter. This is why, describing histamine, Dale could say that it was a product of "organic auto-pharmacology."[16] But it is the same in the pharmacopoeia of the living as in the pharmacopoeia of the scientist: depending on the case, the duration, and the amount, remedies can also be poisons. In short, non-Hippocratic medicine is not anti-Hippocratic any more than non-Euclidean geometry is anti-Euclidian. Nature's capacity to cure is not denied by a treatment that governs it by integrating it—it is located in its proper place, or more exactly, it is comprised within the treatment's limits. Hippocratism noted that the forces of nature are limited, something that led Asclepiades to call expectant medicine a meditation on death. By diverting these forces, non-Hippocratic medicine can move back these limits. Today, ignorance would consist in not asking of nature what is not its own. The medical art is the dialectic of nature.

In our historical outline of the revolution in pathology, we intentionally preferred the work of Loewi over so many others—such as James Reilly or Hans Selye. Loewi's works were resumed and extended at Harvard by Walter B. Cannon and his school. It was Cannon who broadened interest in physiological research on the autonomic nervous system by demonstrating its role in the homeostatic regulation of fundamental biological functions such as circulation, breathing, and thermogenesis.[17] It is Cannon who, after Claude Bernard, presented the functions of regulation in terms of "a modern interpretation of the *vis medicatrix*,"[18] an interpretation that generated optimism as regards the doctor's cooperation with nature—cooperation only in the sense that "nature herself is working with the curative agencies which he [the physician] applies."[19]

It is clear why, ever since physiological science allowed the doctor to count on the existence of protective mechanisms of organic stability, it has become possible for doctors to cease invoking Nature as the providence of Life. Yet it is also clear why, until now, this interpretation, however often it was contested as metaphysical by many positivist minds, could incessantly, and for no less vigorous minds, be authorized, in theory as well as in practice, by the attentive and reliable observation of certain reactions and performances of an organism in a state of disease. If the human organism comprises apparatuses [*dispositifs*] that insure it against risks inherent in its relationship with its milieu, what is so astonishing if these apparatuses function, and what is so foolish if men—patients or doctors—admire their manifest effects?

The review of topics and theses inspired by the practical confidence—for want of theoretical lucidity—in the curative power of nature requires the study of a considerable medico-philosophical literature. The best presentation of this literature is offered in Max Neuburger's work *The Doctrine of the Healing Power of Nature Throughout the Course of Time* (1926).[20] In her recent doctoral thesis, titled "Le médecin de soi-même" ("One's Own Doctor," 1972), Evelyne Aziza-Shuster studied the part of this literature that relates to what one could call "Tiberius's prescription."[21] Tacitus, Suetonius, Pliny the Elder, and Plutarch preserved for posterity the example and exhortation of the emperor Tiberius: past the age of thirty, any man must be able to be his own doctor. After thirty years: that is to say, once enough experience in matters of food, hygiene, and way of life has made it possible for the individual's judgment to separate, among the effects of instinctive (and thus natural) choices, between optimal satisfactions, on the one hand, and on the other, the consequences of docile submission to the rules of an art that is badly founded or intentionally misleading. Who would be surprised to find Montaigne citing Tiberius in order to authorize himself to follow only his appetites, in health and disease, and to make "all the conclusions of medicine give way" to his pleasure?[22] But to find Descartes, after flattering himself for founding an infallible medicine on a science of the living body as solidly demonstrated as mechanics, propose to Burman, as a measure of health, the instinctive discernment of the useful from the harmful that is specific to animals or a confidence in the capacity recognized in Nature of restoring itself, beginning from a state in which "she knows better than the doctor, who is on the outside."[23] What support for the thesis that you are your own doctor! To our knowledge, the first work to bear this title was that of the surgeon Jean Devaux (1649–1729), *Le Médecin de soi-même, ou L'art de conserver la santé par l'instinct* (*One's Own Doctor, or the Art of Remaining Healthy by Instinct*, 1682).[24] A surgeon's diatribe against physicians, the work is also an anti-Cartesian justification of Cartesian naturalism, something about which Devaux is obviously ignorant. He wants to show both that man has instinct, just like any animal, and that instinct in animals is not a mechanism, but knowledge by images. While the work of the Englishman John Archer (who died in 1684), *Every Man His Own Doctor* (1673),[25] preceded that by Devaux, it does not belong to the genre of demonstrative argumentation—it is a self-advertising work by a well-known charlatan. In fact, the medical literature inspired by naturalism remained, remains, and will undoubtedly remain for a long time divided between two intentions or two motivations: a sincere reaction of compensation in times of crisis in therapeutics and an astute utilization of the distress of the patient for the sale of some snake oil, even in print form.

In the eighteenth century, Georg Ernst Stahl chose the title *De medicina sine medico* (*On Medicine Without a Doctor*, 1707) and that of *De autocratia naturae* (*On the Autocracy of Nature*, 1696) to present his excitement over the fortunate contrast, in an organism condemned by its chemical composition to rapid decay, between the propensity toward illness and the scarcity of diseases, which he thought an effect of nature's promptness to restore the animal economy, thanks to the spontaneity of the vital tonic movement.[26] Friedrich Hoffmann, Stahl's rival in Halle, in turn used *De medico sui ipsius* (*Doctor of Oneself*, 1768) to title his mechanist theory of the living body as the rational support of a practice faithful to Hippocratic principles.[27] It is under the same title, *Medicus sui ipsius* (1768), that Linnaeus put forward, more explicitly than so many others since Galen, principles for conducting a life regulated by the use of the six non-natural things[28] as instruments of health and foundations of hygiene.[29] We thus see that the greatest names of medicine and natural history in the eighteenth century did not hesitate to support with their authority a thesis that as a result of the retreat of skepticism or of a therapeutic nihilism would gradually be condemned to survive only in publications of dispute, charlatanry, or retrograde popularization.

In the nineteenth century, the works bearing the same title are works of domestic medicine, of popular medicine, or of philanthropic intention: Handbooks of Health, Friends of Health, Conservators of Health, Regulators of Health, Medicines without a Doctor, Doctors without Medicine, etc. In the absence of an exhaustive census, Evelyne Aziza-Shuster has drawn a systematic tableau of this in her aforecited thesis.

What forced the topic of healing nature to take refuge in popular literature lies in the conjunction of anatomical pathology and new techniques of clinical exploration, such as percussion or auscultation: it is the discovery by Austrian and French doctors in the nascent nineteenth century of phenomena of nature's spontaneous silence.[30] In Vienna and in Paris in the early 1800s, the new clinical practice observes that nature speaks only if one interrogates it well.

From the moment when medicine founds its diagnosis no longer on the examination of spontaneous symptoms, but on the examination of provoked signs, the patient's and the doctor's respective relations to nature are turned upside down.[31] Because he cannot make out the difference between signs and symptoms on his own, the sick man is led to believe that any conduct that is regulated exclusively by symptoms is natural. For his part, the doctor now knows that he cannot accept all that nature says and the way in which it says it without first using his art to force nature to express itself; and for this reason, he is thus led to defy nature—not only in what it says, but also in what it does. While in his *agrégation* thesis, *De l'expectation en médecine* (*Of Expectation in Medicine*, 1857), Jean-Martin Charcot introduces subtleties into his argument about expectation in order to retain some credit for naturalism and humorism,[32] Emile Littré, faithful to the positivist teaching in which action is founded on science, takes up Tiberius's dictum only in order to refute it: Littré recalls to the patient the obligation to not put his trust in his own sense, but to appeal to the man capable of knowing for him what he is himself unaware of—that is, to the doctor.[33] It is no longer a question of supplanting medicine by hygiene. There is no hygiene without a doctor.

Physiology has justified some of the intuitions of naturist ancient medicine through its progressive discovery of organic mechanisms of self-regulation and stabilization, mechanisms whose explanation is sought today in models of active reaction, in other words, of feedback.

At the same time, the therapeutics of infectious diseases, ever since Louis Pasteur, Robert Koch, and their pupils, has legitimized—until recently without proof, if not without argument—the attribution to the organism of an innate capacity of antitoxic defense. Here, to include is to surpass: the guidance of spontaneous immunization by way of immunological techniques excites a curative reaction not by duping the body but by introducing a lesser evil [*mal*], a benevolent evil, which leads the organism to react more promptly than usual so as to outstrip the onset of more serious, imminent harm [*mal*]. Moreover, and better yet, one can transform an animal organism into the permanent producer of transmissible natural remedies. Wilhelm Roux, Emil Adolf von Behring, Paul Ehrlich: three great craftsmen of the taming of a "wild" [*sauvage*] healing nature.[34] Through Ehrlich's ingenuity, modern chemotherapy was born in the systematic study of the cellular modes of reaction, which were baffling in their partiality, because the spontaneous production of antibodies that is recuperated in the techniques of vaccination and serotherapy was not observable in the case of the protozoa.[35]

Contemporary medicine cannot honor Hippocrates better than by ceasing to reclaim him for itself; it cannot celebrate the approximate accuracy of his conception of the organism except by refusing his practice of observation and a wait-and-see approach [*expectation*]. It is not prudent to wait for nature to declare itself once we have verified that to know its resources, one most mobilize them by alerting them. To act is to activate, as much to reveal as to cure.

One can thus continue—even in the age of industrial pharmacodynamics, of the imperialism of the biological laboratory, of the electronic treatment of diagnostic information—to speak of "nature" when designating the initial fact of existence of self-regulating living systems whose dynamics are inscribed in a genetic code. And in all rigor, we must tolerate that for the sick, confidence in the power of nature could assume the form of mythical thought: the myth of origin, the myth of the anteriority of life to culture. One can psychoanalyze the myth and find the face of the Mother in the figure of Nature. No matter—on the contrary. Until a new order comes, the biological order is primordial vis-à-vis the technological order. It was a heterodox psychoanalyst, Georg Groddeck, who worked out the first concepts of what had to be called "psychosomatic medicine" by developing the naturist teaching of Ernst Schweninger, Bismarck's personal doctor. Groddeck entitled the book that he dedicated to him in 1913 *Nasamecu*.[36] *NAtura SAnat, MEdicus CUrat*; Nature heals, the doctor cures.

TWO

Diseases

At the beginning of his *Essays on Painting*, Diderot writes: "Nature does nothing incorrectly. Every form, beautiful or ugly, has its reason; and of all the beings that exist, there is not one that is not as it must be."[1] One can imagine an *Essays on Medicine* that would begin with: "Nature does nothing arbitrarily. Disease, like health, has its reason, and among all the living beings, there is not one whose state is not as it must be."[2] This kind of prologue could not concern all populations at all times. Over centuries and in many places, illness was either a possession by some "malignant" being against which only a miracle worker could triumph or a punishment inflicted by a supernatural power on the deviant or the impure. There is no need to travel to the Far East for examples of this. Already in the Old Testament, (Leviticus 13 and 14) leprosy is considered and rejected as an impurity, and lepers are being chased out of communities. In Greece, the first forms of cure and therapeutics are of a religious order. Asklepios, son of Apollo, is the healing god; his priests are his executants. It is in Asklepios's temples that patients are received, examined, and treated according to rites from which the serpent and the cock still survive today as symbolic participants.

One can speak with reason of "Greek medicine" only from the Hippocratic period onward—that is to say, from the moment when diseases came to be treated as bodily disorders about which it is possible to construct a communicable discourse concerning their symptoms, their supposed causes, their probable course, and the behavior that must be observed if one is to correct the disorder that all these indicate. It has always been noted that this medicine, for which Hippocrates' *Aphorisms*[3] are a sort of gospel, is contemporary with the first investigations meriting the name of science and with the rise of philosophical thought. Plato's dialogue *Phaedrus* contains a tribute to Hippocrates, whose method is declared to conform to "right reason."[4]

Nevertheless, one would not accept that such medical practice, however secular and reasoned, could be termed "scientific" in the modern sense of the term. Contemporary medicine is founded, with an efficacy we cannot but appreciate, on the progressive dissociation of disease and the sick person, seeking to characterize the sick person by the disease, rather than identify a disease on the basis of the bundle of symptoms spontaneously presented by the patient. Disease points us to medicine rather than to evil. When a doctor speaks of Basedow's disease (that is, of *goiter exophthalmia*), he designates a state of endocrine dysfunction whose presentation of symptoms, etiological diagnosis, prognosis, and therapeutic decision making are all supported by a succession of clinical and experimental studies and laboratory tests, in the course of which patients are treated not like subjects of their disease, but like objects.

Plague, cancer, shingles, leukemia, asthma, and diabetes are also species of organic disorder that are felt by the living as a harm, an evil [*un mal*]. Disease is the risk of the living as such—risk as much for the animal or the vegetal as it is for the human. Yet in this last case, and by contrast with the risk that is born from a resolution to act, the risk that is born with birth is quite often inevitable. Suffering, the restriction of chosen or required habitual activity, organic deterioration, and mental decline are all constitutive of a state of harm, but are not by themselves specific attributes of what the physician of today identifies as disease, even at the moment when he endeavors to put an end to this harm or to attenuate it. Nevertheless, the patient-disease relation cannot be one of complete discordance. In contemporary societies, where medicine endeavors to become a science of diseases, the institutions of public health and the popularization of medical knowledge have the effect that, for the patient, living his disease means talking and hearing talk in clichés or stereotypes; living his disease also means implicitly valorizing the results of a knowledge whose progress is due in part to putting the patient between brackets, however much he may be the avowed subject of medical concern.

The current understanding of somatic diseases is the doubtlessly provisional outcome of a succession of crises and inventions in medical knowledge, of progress concerning the practices of examination and the analysis of their results. The effect of this outcome is to force doctors to shift the seat and revise the structure of the pathogenic agent and, further, to change the target of reparative intervention. Correlatively, the sites of observation and analysis of suspected organic structures have been displaced on the basis of apparatuses and techniques proper to them or imposed on them. It is in this way that diseases have been successively localized in the organism, the organ, the tissue, the cell, the gene, the enzyme. And work to identify them has taken place successively in the autopsy room, then in laboratories where physical examinations (optical, electrical, radiological, scannographic, echographic), then chemical or biochemical examinations are conducted.

The increasingly close relation between medicine and biology has permitted us, thanks to a more exact knowledge of the laws of heredity, to distinguish among diseases those that are hereditary (depending on the constitution of the genome), those that are inborn (depending on the circumstances of intrauterine life), and those that are, properly speaking, occasional (that is, results of the relations of the individual to the ecological milieu as well as to the social group). This may concern individual accidents, such as pneumonia, or collective ones, such as the flu or typhus, that is, diseases known as infectious, whose own birth, life, and death have been the subject of studies by Charles Nicolle.[5] Undoubtedly, in the history of societies and civilizations, these diseases should be considered natural phenomena characterized by epoch, sites of appearance, diffusion, and extinction. But even though we have known, at least since the end of the nineteenth century, their determining causes (microbes, bacilli, viruses), on the one hand, and, on the other, their vectors (the flea on the rat for the plague, the *Aedes aegypti* mosquito for yellow fever), nevertheless, the historian of these diseases cannot neglect to attend to the reasons of their geographical distribution or to the form of social relations proper to affected populations. Moreover, the contemporary collective struggle, as a matter of public hygiene, is one of the determining factors of the picture we have of these diseases, of the way in which they evolve, in terms of their symptoms and course, under the effect of the means of struggle they provoke. One can hardly ignore that the consequence of the generalized practice of vaccination has been the appearance of varieties of microbes that are more resistant to vaccines. And this is but one of the aspects of a purposeful inversion that turns the multiplication and improved efficiency of medical and surgical activities—in industrial societies with a highly technical sanitary protection—into a growing risk of failures in the internal biological system of resistance to diseases.

In the human environment, initially natural and then gradually more artificial, there is nothing that could not be considered a source of danger to someone or other, to the extent that the concept of the human covers over with a false appearance of specific identity individual organisms whose ancestry has endowed them with different powers of resistance to attack. What are called "innate errors of metabolism" or "hereditary biological anomalies" render certain individuals or populations sensitive and susceptible to situations or objects of paradoxical harmfulness. For someone from the Mediterranean, eating raw fava beans amounts to poisoning oneself, because genetic inheritance has deprived one of a certain amylase. And yet the same enzymatic deficit has accorded certain African populations with increased resistance to malaria.[6] At present, there are many cases in which to identify a disease, it is necessary to learn not to seek access to it by including the patient. The enzymologist can see states of real, albeit latent and provisionally tolerated disease that the clinician, who observes spontaneous or induced signs as they appear on the register of the organism or the organ, would ignore.

The progressive elimination from the knowledge of diseases of reference to the living conditions of patients is not only the effect of the colonization of medicine by the general and applied sciences since the early nineteenth century. It is also an effect of the interest (in every sense of the term) of industrial society since that period in the health of working populations, or, as some put it, in the human component of productive forces. The monitoring and amelioration of living conditions have been the object of measurements and regulations developed by political authorities thanks to the behest and clarifications of hygienists. Medicine and politics thus encountered one another in a new approach to diseases, as we find convincingly illustrated by the structures and practices of hospitalization. Already in eighteenth-century France, particularly during the Revolution, steps were taken to replace hospices, shelters that provided relief for sick patients who were often abandoned, with hospitals, spaces of analysis and monitoring of classified diseases built to function as "curing machines" (to use a phrase coined by Jacques Tenon).[7] Treatment of diseases in hospitals under a regulated social structure contributed to their deindividualization at the same time that the progressively more artificial analysis of their conditions of appearance separated their reality from their initial clinical representation.

The corollary of this theoretical disengagement was a mutation in the medical profession and in the approach to diseases. The doctor-therapist concerned with all aspects of illness and currently known as "general practitioner" witnessed declining prestige and authority in favor of specialist physicians, engineers who take apart the organism like machinery. These are still doctors in terms of function, though no longer by reference to the lay idea of a doctor, because consultation now consists of a combination between a computer search through semiological and etiological databanks and a formulation of a probability diagnostic supported by the evaluation of statistical information. We have to note on this subject that the emergence of the study of diseases from a statistical point of view concerned with their appearance, social context, and evolution is precisely contemporary with the anatomical-clinical revolution in hospitals in Austria, England, and France at the beginning of the nineteenth century. In short, one cannot deny the existence of a social, hence political, component in the invention of theoretical practices that are currently effective in the knowledge of diseases.

Should the introduction of a sociopolitical point of view into the history of medicine be confined to the investigation of causes of a change in medical knowledge and direction? Shouldn't one also recognize causalities of a sociological order in the appearance and course of diseases themselves? We have recently seen independent trade unionists denounce capitalism's diseases—which amounts to seeing in disease the organic index of class relations in capitalist societies. There was a time when one spoke of "diseases of misery," that is, disorders that would emerge in certain sections of the population out of vitamin deficiencies resulting from malnutrition. In fact, the first medical discipline preoccupied with these types of questions was hygiene. In the introduction to his *Élements d'hygiène* (*Elements of Hygiene,* 1797), Étienne Tourtelle insisted on the pathogenic incidence of the density of populations in modern towns.[8] In England, as in France, in the first third of the nineteenth century, many investigations were carried out on the health of workers in various branches of industry. In 1840, Villermé published his famous *Table of the Physical and Moral State of Workers Employed in the Manufacture of Cotton, Wool, and Silk.*[9] Treatises on occupational hygiene were numerous in France in the nineteenth century. Nevertheless, whatever importance in the increase of pathological situations we should accord to patterns of life and their relations to working conditions (for example, the role of muscular exhaustion or the deregulation of functional rhythms), it is improper to confuse the social genesis of diseases with the diseases themselves. The clinical picture of diseases such as stomach ulcers and pulmonary tuberculosis is unaffected by them being the effect of individual or collective situations of distress. Even if the work of the clockmaker or the duties of the schoolboy expose defects in vision much more easily than shepherding sheep would, we would nevertheless not go so far as to say that ophthalmology patients are social facts. There are, however, cases in which the census and evaluation of disease factors can take into account the social status of patients and their representations of it. To use a vocabulary made fashionable by the works of Hans Selye, let us say that one could include among the pathogenic forms of *stress* (that is, among nonspecific responses) the individual's perception of the extent of his insertion into a professional or cultural hierarchy.[10] The fact of living with a disease, like experiencing a forfeiture or a devaluation—and not just as a form of suffering or behavioral limitation—is one that we must consider among the components of the disease itself. Here we find ourselves on the nebulous border between somatic medicine and psychosomatic medicine. This is a border haunted by psychoanalysis, because what is in question is as much the unconscious as the techniques devoted to making it speak in order that we may know how to respond.

From the perspective of medical psychosociology—which today is quite in vogue—one could even regard illness as the poorly studied readiness to oblige of a patient placed in situations like that of the victim or the condemned. Without going so far as to consider such recastings of myths as ethnology's revenge on biology in matters of explaining diseases, we can still see in them the distant effect of a resistance to the extremism of medical theories indentured to Pasteurism or, more recently, exalted by the successes of molecular biochemistry. In any case, we must recognize that current methods of identifying diseases and therapy are due to the successes of immunology, rather than to thaumaturgies inspired by psychosociology. Immunology is a biochemical discipline based on medical experience. Its most remarkable aspect is to have founded the singularity of the patient at the level of the molecular structure of the organism's cells—a singularity that medical personalism and the propaganda of the "guerilla armies of medicine" [*francs-tireurs de la médecine*] celebrate against the impersonal essence of the disease.[11] That conception of disease retained some vestiges of the old theory of "morbid species" elaborated in the seventeenth century by Thomas Sydenham.[12] The conceptual revolution concerning diseases was the identification of what came to be called the immune system, that is, a totalizing structure of responses to aggressive antigens by way of the production of specific antibodies. The collaboration of clinic and laboratory in immunological research is perhaps still fragile, but it has introduced the reference to biological individuality in the representation of disease. The opposition, often sharp in the nineteenth century, between medical and scientific conceptions of disease has been replaced by a common hope to find someday, through molecular biology, an effective response to diseases that are nowadays burdened with phantasms of distress: cancer and AIDS. In effect, one cannot dissociate the existence and behavior of diseases from the changes occurring in the epistemological status of medicine. The best example of this is the recent extinction of smallpox as a result of the preventive vaccination measures derived from Pasteurian bacteriology. One cannot treat disease the way one treats meteorological phenomena, even as, in this last case too, the activity of *Homo faber* on the surface of the earth affects climates.

Whatever the interest of a study of the varieties, history, and effects of a disease, it should not eclipse the interest of attempts to comprehend the role and sense of disease in human experience. Diseases are crises in the growth toward the adult form and structure of the organs and in the maturation of functions of internal self-conservation and adaptation to external demands. They are also crises in the effort undertaken to live up to a model of selected or imposed activities and, in the best cases, to defend the values of or reasons for living. Diseases are a ransom eventually to be paid by men made to live without having asked for it, who must learn that they necessarily tend, from their very first day, toward an unforeseeable and inescapable end. This end can be precipitated by diseases, whether violent or merely responsible for a diminution in the capacity to resist other diseases. Conversely, certain diseases can, after recovery, confer on the organism a capacity to resist others. To grow old, to last, if not unscathed, then at least toughened, can thus also be the benefit of having been sick.

As a universal biological fact and a singular existential test of man, the existence of disease provokes an interrogation of the precariousness of organic structures that up to now has not found a convincing response. None among the living are strictly speaking complete [*achevé*]. Whether one calls this evolution or not, whatever explanation one gives it, the historical succession of organisms, on the basis of what is now known as prebiotic chemical evolution, is a succession of claimants not powerful enough to be living beings that are more than merely *viable*—that is to say, beings that are fit to live, but lack any guarantee of succeeding totally in doing so. Death is in life—and disease is the sign of this. The meditation on the, strictly speaking, mortifying experience of disease is often more poignantly expressed in poems than in sermons. But it was a physician particularly sensitive to the suffering of others and himself suffering from cancer who succeeded in achieving in simplicity the profundity of pathos. In letters addressed to Lou Andréas-Salome, Freud wrote: "I have stood the foul realities pretty well, but I don't bear well the thought of the possibilities before me, and I cannot get used to the idea of a life under sentence."[13] And elsewhere: "A crust of indifference is slowly creeping up around me; a fact I state without complaining. It is a natural development, a way of beginning to grow inorganic."[14] Between revolt (incited by the idea of a life under a death sentence) and resignation (the acceptance of a return to the inorganic), disease has done its work. Work [*travail*], according to etymology, is torment and torture [*tourment et torture*]. Torture is the suffering inflicted to obtain a revelation. Diseases are instruments of life by which the living (when it is man we're talking about) see themselves forced to avow their mortality.[15]

↛

The constitution of diseases is patterned after the constitution of animals. They have their destiny, limited from their birth, and their days. He who tries to cut them short imperiously by force, in the midst of their course, prolongs and multiplies them, and stimulates them instead of appeasing them. I agree with Crantor, that we must neither obstinately and heedlessly oppose evils nor weakly succumb to them, but give way to them naturally, according to their condition and our own. We should give free passage to diseases; and I find that they do not stay long with me, who let them go ahead; and some of those that are considered most stubborn and tenacious, I have shaken off by their own decadence, without help and without art, and against the rules of medicine.[16]

Any morbid disease, whatever it is, always corresponds to a physiological function that has been diverted and changed in some way; differently put, all disease has a normal corresponding function of which it is a disordered, exaggerated, weakened, or annulled expression. If we cannot today explain all phenomena of disease, this is because physiology is not yet sufficiently advanced and there is still a whole crowd of normal functions unknown to us. But this way of seeing is not the one usual to the majority of pathologists. Most of them, in effect, accord to disease a sort of individuality, an extraphysiological personality that would attach itself to the organism and would have its own laws, independent of physiological laws. One author even defined disease as "a function that leads to death," by contrast to a normal function that maintains life. I do not need to say that this definition of disease seems to me pure fantasy. All functions have as their object the maintenance of life and constantly tend to reestablish the physiological state when it is disordered. This tendency persists in all morbid states, and this is the one that constituted the healing force of nature for Hippocrates.[17]

In the eighteenth century, disease was both nature and counter-nature, since it possessed an ordered essence, but it was of its essence to compromise natural life. From Bichat onwards, disease was to play the same dual role, but between life and death. Let us be clear about this: an experience devoid of both age and memory knew, well before the advent of pathological anatomy, the way that led from health to disease, and from disease to death.[18]

THREE

Health: Popular Concept and Philosophical Question

"Who among us did not speak of what is healthy and what is harmful before the arrival of Hippocrates?" This is how Epictetus, in his *Discourses*, argues for the popular pertinence of an a priori notion of the healthy and of health—a health whose relation to objects or behavior, moreover, he considers to be uncertain.[1] If we were to allow that such a definition of health might be possible without reference to some explicit knowledge, where would we seek its foundation?

It would be inappropriate, here in Strasbourg,[2] to offer you some reflections on health without recalling the definition suggested a half century ago by a famous surgeon and professor in the Faculty of Medicine from 1925 to 1940: "Health is life lived in the silence of the organs."[3] It was perhaps thanks to conversations between colleagues at the Collège de France that Paul Valéry came to echo René Leriche, writing that "health is the state in which necessary functions are achieved imperceptibly or with pleasure."[4] Earlier, Charles Daremberg, in his collection of articles *La médecine, histoire et doctrines* (*Medicine: Its History and Doctrines*, 1865), had written that "in health, one does not feel the movements of life; all functions are accomplished in silence."⁵ And after Leriche and Valéry, Henri Michaux also likened health to silence, though he did so in the negative: "Just as the body (its organs and its functions) has been mainly known and revealed not by the prowess of the strong, but by the disorders of the sick, the weak, the infirm, and the wounded (health being silent, the source of an immensely erroneous impression that everything goes along all by itself), it is the disturbances of the spirit, its dysfunctions, that will be my teachers."⁶ Well before all these, and perhaps more subtly than any of them, Diderot had written in his *Letter on the Deaf and Dumb* (1751): "When we are well, no part of the body informs us of its existence; if by some pain it informs us of itself, it is then certain that we are not doing well; and even if we feel pleasure, it is not always certain that we are doing better."⁷

Health was a frequent philosophical topic during the classical period and the Enlightenment, and it was almost always approached in the same way —by reference to disease, whose absence was generally held to be the equivalent of health. Thus, for example, Leibniz, while discussing Pierre Bayle's theses on good and evil in *Theodicy* (1710), writes: "But does physical good lie solely in pleasure? Monsieur Bayle appears to be of this opinion; but I consider that it lies also in a middle state, such as that of health. One is well enough when one has no illness; it is a degree of wisdom to have no folly."⁸ Leibniz adds: "Monsieur Bayle would wish almost to set aside the consideration of health; he likens it to rarefied bodies, which are scarcely felt, like air, for example; he likens pain to bodies that have much density and much weight in slight volume. But pain itself makes us aware of the importance of health when we are bereft of it."⁹

Kant is noteworthy among the philosophers who paid great attention to the question of health. Strengthened by the successes and failures of his personal art of living (of which Andreas Christoph Wasianski wrote a long account in 1804),¹⁰ Kant addressed the question of health in the third section of *The Conflict of the Faculties* (1798). As for health, he says, we find ourselves in an embarrassing position: "[One] can *feel* well (to judge by his comfortable feeling of vitality), but he can never *know* that he is healthy. . . . Hence if he does not feel ill, he is entitled to express his well-being only by saying that he is *apparently* in good health."¹¹ In spite of their apparent simplicity, these remarks by Kant are important, because they make health an object outside the field of knowledge. We can put Kant's statement more strongly: there is no science of health. Let us accept this for the moment. Health is not a scientific concept; it is a popular concept. Which is not to say that it is trivial, but simply common, within everyone's reach.

It seems to me that at the head of this series of philosophers—Leibniz, Diderot, Kant—we should place Descartes. His conception of health is especially important, since he is the inventor of the mechanist conception of organic functions. This philosopher, who acted as his own doctor, seems to me, by associating health and truth in his tribute to silent values, to have posed a question that up to now has been only poorly understood. In a letter to Chanut (March 31, 1649), he wrote: "Though health is greatest among all those of our goods which relate to the body, it is however that which we reflect on the least and which we taste the least. The knowledge of truth is like the health of the soul: once a man possesses it, he doesn't think of it anymore."[12]

How is it that no one ever thought of reversing this comparison, that no one ever asked whether health were the truth of the body? Truth is not only a logical value specific to the exercise of judgment. There is another sense of truth—for which we need not turn to Heidegger.[13] In Emile Littré's *Dictionnaire de la langue française* (*Dictionary of the French Language*), the article titled "Truth" starts as follows: "Quality by which things appear such as they are."[14] *Verus*, "true" [*vrai*], is used in Latin in the senses of "real" and "regular" or "correct." As for *sanus*, "healthy" [*sain*], it is the descendant of the Greek *saos*, which also has two senses: "intact or preserved," and "infallible or sure." Hence the expression "safe and sound." In his *Histoire des expressions populaires relatives à l'anatomie, à la physiologie et à la médecine* (*History of Popular Expressions Related to Anatomy, Physiology, and Medicine*, 1888), Édouard Brissaud employs a saying that one can interpret as a kind of popular recognition of the alliance between health and truth: "As foolish as a sick athlete."[15] "Foolish" [*sot*] here suggests at once "stupid" and "deceived." The athletic habitus points to a maximum possession of physical means, the equation of ambitions with capacities. A sick athlete is a confession that his body has been rendered false.

But it is an author in the German language, subtler in his choice of references than any collector of sayings, who offers unexpected support to what I would call a thesis waiting for an author. This is Friedrich Nietzsche. It is not easy, even after so many commentators—notably Andler,[16] Bertram,[17]

Jaspers,[18] and Löwith[19]—to determine the sense and scope of Nietzsche's many texts relating to disease and health. In *The Will to Power*, following Claude Bernard, Nietzsche at times believes in the homogeneity of health and disease; at others, he celebrates the "great health" capable of absorbing and overcoming morbid tendencies.[20] In *The Gay Science*, this great health is the power to put all values and desires to the test.[21] In *The Antichrist*, Christianity is denounced for its incorporation of the instinctive rancor of the sick toward the healthy, for its loathing of "all that is straight, proud, and superb."[22] Retain: "straight" [*droit*]. In *Thus Spoke Zarathustra*, we find the same rectitude of the body opposed to the morbid preachers of the beyond: "More honestly and more purely speaks the healthy body, the perfect and perpendicular body and it speaks of the meaning of the earth."[23] Is it superfluous to recall here that in Chinese mythology, the square is the symbol of the earth, whose form is square, whose divisions are square? Health thus summarizes, for Nietzsche, straightness, reliability, competence. Further on, "the body is a great reason, a peace, a herd and a shepherd." Finally, "there is more reason in your body than in your best wisdom."[24]

By the time Nietzsche wrote this in 1884, physiologists had experimentally established the existence of apparatuses and functions of organic regulation. But it is not very likely that the great English physiologist Ernest Henry Starling was thinking of Nietzsche when he titled his discourse on regulation and homeostasis "The Wisdom of the Body" (1923), a title also taken up by Walter B. Cannon in 1932.[25] As far as health is concerned, Starling (inventor in 1905 of the term *hormone*) published a treatise, *Principles of Human Physiology* (1912)—later revised by Lovatt Evans—whose index does not even contain the entry "health."[26] Similarly, "health" [*santé*] does not appear in the index of Kayser's *Physiologie* (1963).[27] In each of these treatises, by contrast, the index contains "homeostasis," "regulation," and "stress." Should we see here a new argument for refusing the concept of health any scientific status?

Can we, must we, say that the functions of the organism are objects of science, but that what Bernard called "the harmonic relations of economic functions" are not? Besides, Bernard expressly said that "in physiology, there exist only conditions specific to each phenomenon, which we should determine exactly, without losing ourselves in hallucinations on life, death, health, disease, and other entities of the same kind." Which did not prevent him from later using the expression "organism in a state of health."[28]

Nonetheless, in its general introduction, Starling's *Principles* contains a remark that may seem minor, but that I believe is important enough to highlight. It indicates, for students, that the term "mechanism," often used to depict how an organic function operates, should not be given too much weight. ("This rather overworked word need not be taken too seriously.")[29]

We are comforted by this refusal to reduce health to a necessary effect of mechanical relations. Health, the body's truth, does not arise out of an explanation of theorems. There is no health for a mechanism. Besides, Descartes himself, in the Sixth Meditation, while denying an ontological difference between a working clock and a clock that is broken, teaches us that there is an ontological difference between the latter and a hydropic man, that is, an organism driven by thirst to drink without constraint. When drinking is harmful, says Descartes, it is an error of nature to be thirsty.[30] By health, Descartes understands "aliquid . . . quod revera in rebus reperitur, ac proinde nonnihil habet veritatis"—"something that is really to be found in things themselves; in this sense, it therefore contains something of the truth."[31] For a machine, the operative state is not health, and disorder is not a disease. No one has said this as profoundly as Raymond Ruyer in his *Paradoxes de la conscience (Paradoxes of Consciousness)*. It is enough to quote, from among several passages, one that concerns the cybernetic vicious circle. It is absurd to conceive the living organism as a regulated machine, since ultimately—and whatever the intermediary stages may be—"the regulated machine is always in the service of a regulation or a conscious organic selection," but "by definition, a natural regulation can be only . . . self-regulation without a machine."[32]

There is no disease of the machine, just as there is no death of the machine. Auguste Villiers de l'Isle-Adam, whose merits are the subject of vivid debate but who is nonetheless credited with having inspired Mallarmé, imagined in *Tomorrow's Eve* (1886) an Edison who invents the electromagnetic means to simulate the functions (speech included) of a living human being. His android, Hadaly, is a woman-machine who can say "I," but who knows herself to be nonliving, since no one addresses her as "you," and who declares in the end: "I, who extinguish myself, will not be returned to Nothingness. . . . I am the obscure being whose death is not worth a single mournful memory. My unfortunate bosom is not worthy even to be called sterile. If only I could live, if only I possessed life. . . . To only be able to die."[33]

The living body is thus the singular being whose health expresses the quality of the forces that constitute it: it must live with the tasks imposed on it, and it must live exposed to an environment that it does not initially choose. The living human body is the totality of the powers of a being that has the capacity to evaluate and represent to itself these powers, their exercise, and their limits.

This body is at once a given and a product. Its health is at once a state and an order.

The body is a given to the extent that it is a genotype, a necessary and also singular effect of the components of genetic inheritance. In this context, the truth of its presence in the world is not unconditional. It sometimes follows from errors of genetic coding, which, depending on the milieu, may or may not determine pathological effects. The untruth of the body can be manifest or latent.

The body is a product to the extent that its activity of insertion in a specific milieu, its (selected or imposed) way of life, sport, or work, contributes to fashioning its phenotype, that is, to modifying its morphological structure and to beginning to individualize its capacities. Here, the discourse of hygiene finds its occasion and justification: it is a traditional medical discipline, recently recovered and travestied by the socio-politico-medical ambition of regulating the lives of individuals.

Ever since "health" was first said to belong to man as a participant in a social or professional community, its existential meaning has been occulted by the demands of accounting. Tissot had not quite arrived at this when he published his *L'avis du peuple sur la santé* (*The People's Opinion on Health*, 1761) and *De la santé des gens de lettres* (*On the Health of Men of Letters*, 1768).[34] But health was starting to lose its meaning as truth and to receive a meaning as facticity. It was becoming the object of a calculation. Ever since, we have known this as the checkup. It is worth recalling that it was here in Strasbourg that Étienne Tourtelle, professor at the École Spéciale de Médecine, published in 1797 his *Elements of Hygiene*.[35] The historical broadening of the space in which administrative control over the health of individuals is exerted has today led to a World Health Organization, which could not delimit its field of intervention without publishing its own definition of health: "Health is a state of complete physical, moral, and social well-being and not merely the absence of disease or infirmity."[36]

Health, as the state of a *given* body, is proof—by virtue of the fact that this living body is possible, since it exists—that it has not been congenitally altered. Its truth is security. Is it not astonishing, then, that sometimes, and quite naturally, we speak of "fragile" or "precarious" health and even of "bad" health? Bad health involves a shrinking of the margins of organic security, a limitation of the power to tolerate and compensate for the aggressions of the environment. In a famous conversation in Amsterdam in 1648, the young Burman used the fact of disease to dispute with Descartes—trusting in the rectitude of the constitution of the body to control and prolong human life. Descartes' response might be surprising. He said that nature remains the same, that it seems to plunge man into diseases so that, by surmounting them, he can become more able-bodied. Descartes obviously could not anticipate Pasteur. But is vaccination not precisely the artifice of a calculated infection that allows the organism to defend itself from an infection in the natural state?

Health, as the expression of the *produced* body, the body as product, is lived assurance—in the double sense of insurance against risk and the audacity to run this risk. It is the feeling of a capacity to surpass initial capacities, a capacity to make the body do what initially seemed beyond its means. We rediscover the athlete. A line from Antonin Artaud, which concerns, first of all, human existence under the guise of life, rather than life proper, can serve to help us define health: "We can accept life only on condition of greatness, only if we feel ourselves at the origin of the phenomena, at least of a certain number of them. Without the power of expansion, without a certain domination over things, life is indefensible."[37]

Here we are far from a health measured by apparatuses. Let us call this health free, unconditioned, unaccountable. This free health is not an object for those who believe themselves to be specialists in it. The hygienist endeavors to govern a population—individuals are not his business. "Public health" is a contestable term—"salubrity" would be more appropriate. Very often, it is disease that is public, it is disease that is publicized. The patient calls for help, draws attention; he is dependent. The healthy man who adapts silently to his tasks, who lives the truth of his existence in the relative freedom of his choices, exists in a society that ignores him. Health is not only life lived in the silence of the organs—it is also life lived in the discretion of social relations. If I say that I am well, I block stereotypical questions in advance. If, by contrast, I say that I am unwell, people want to know how and why; they wonder or ask me whether I am registered with social security.[38] Interest in individual organic failure is eventually transformed into interest in the budgetary deficit of an institution.

But, leaving aside the description of the lived situation of health or disease, let me try to justify the proposal that health should be held to be the truth of the body by considering, in a state of exertion, the original expression of the body's position as a unit of life, which is as the foundation of the multiplicity of its own organs. The recent technique of organ harvesting and transplantation does not take anything away from the capacity of a given body to integrate, in a sense by appropriating, a part taken from a whole with a compatible histological structure.

The truth of my body—its very constitution or its authenticity of existence—is not an idea susceptible to representation, just as, according to Malebranche, there is no *idea* of the soul.[39] By contrast, there is an *idea* of the body in general, certainly not visible and readable in God as per Malebranche, but laid out biologically and medically in progressively verified knowledge. This health without idea, at once present and opaque, is what supports and validates, in fact and in the last resort, for me and for the physician, insofar as he is *my* physician, what the idea of the body, that is, medical knowledge of the body, can suggest as an artifice to sustain it. My physician ordinarily accepts from me what I tell him regarding what only I am able to tell him, that is, what my body announces to me through symptoms whose meaning is unclear to me. My physician accepts that I see in him an exegete before accepting him as repairer. The definition of health that includes the link of organic life to pleasure and pain tested as such surreptitiously introduces the concept of a *subjective body* into the definition of a state that medical discourse thinks it can describe in the third person.

In recognizing in the health of the living human body its truth, have we not agreed to follow Descartes on a path where some of our contemporaries have thought they have discovered the trap of ambiguity? This is Michel Henry's claim in his *Philosophy and Phenomenology of the Body* (1965).[40] By contrast, Merleau-Ponty used to credit Descartes with what most reproach him for as an ambiguity. Even before the posthumous text *The Visible and the Invisible*, he had tackled this question in his course *L'union du corps et de l'âme chez Malebranche, Maine de Biran et Bergson* (*The Union of the Body and the Soul in Malebranche, Maine de Biran, and Bergson*, 1947–48) and in his last lecture course, *La Nature*, at the Collège de France in 1960.[41] In a note in *The Visible and the Invisible*, we read: "The Cartesian idea of the human body as human *non-closed*—open inasmuch as governed by thought—is perhaps the most profound idea of the union of the soul and the body."[42] Ultimately, despite his virtuosity and ambition, Merleau-Ponty could do no more than comment on the unsurpassable.[43] The best commentators are those who admit themselves to be such, in recognizing, in the existence of the living human body, a side that is "inaccessible to others, accessible only to its titular holder."[44] Here we again meet Ruyer, for whom the paradoxes of consciousness are paradoxes only with regard to "our having become accustomed to mechanical phenomena on our scale."[45]

Does our attempt to elucidate a concept of health not run the risk of being taken for delirious raving? In asking philosophy to reinforce our proposal that health be treated as a concept on which everyday experience confers the meaning of a permission to live and act according to the well-being of the body, we appear to scorn the discipline that, even from the popular point of view, seems most appropriate to treating our question: medicine. One could object that, throughout time, the body, sensed and perceived as a force—and sometimes also as an obstacle—has had some relation to the body as represented and treated by medical knowledge. This relationship could even become manifest (in nineteenth-century France) in an institution all but forgotten today—the corps of health officials. These watchmen and advisers in matters of health were in fact subdoctors [*sous-médicins*], from whom a lesser degree of knowledge was expected than from doctors proper. They were in the service of the people, notably in the countryside, where life was held to be less sophisticated than in cities. The body as understood by the people has always been indebted to the body as understood by the faculty of medicine. Even today, the body as understood by the people is often a divided body. The diffusion of an ideology of medical specialization often results in the body being lived as if it were a battery of organs. Conversely, behind the professional and ultimately political debate between specialists and generalists, the medical corps puts back in question, timidly and confusedly, its own relation to health. This effort at professional reform echoes, in a sense, the multiplicity of naturalist protests linked to ecological movements and to an ideology opposed to "resourced health." The same man who militated for a society without schools called for an insurrection against what he named "the expropriation of health."[46] This defense and illustration of "natural, private health" in order to discredit "scientifically controlled health," has taken many forms, including the most ridiculous.

But to find inspiration in Cartesian philosophy when trying to define health as the truth of the body—is this also to say that one can go no further, in the self-management of one's health, than to follow the Cartesian precept of using "the ordinary course of life and conversation?"[47] Can this credit, which we grant to a species of naturalism that one could call "theological," be invoked by the adepts of another, antirationalist naturalism? To advocate "natural" health and the return to an original health by rejecting the scleroses that are considered consequences of eruditely controlled behavior—is this the way to return to the truth of the body? But it is one thing to take charge of the subjective body and another to believe oneself charged with liberating this knowledge from the supposedly oppressive tutelage of medicine and, beyond that, of the sciences that medicine applies. The recognition of health as the truth of the body in an ontological sense not only can, but must admit the presence, as on a ledge overlooking it, of truth in the logical sense—that is, that of science. Admittedly, the living body is not an object—but for man, to live is also to know. I go well [*je me porte bien*] insofar as I feel able to take [*porter*] responsibility for my acts, to bring [*porter*] things into existence and to create between them relations that they would not have without me, but without which they would not be what they are. And thus I need to learn to know what they are in order to change them.

In conclusion, I must undoubtedly justify having made health a philosophical question. This justification will be short: I find it in Merleau-Ponty, who wrote in *The Visible and the Invisible* that "philosophy is the set of questions wherein he who questions is himself implicated in the question."[48]

FOUR

Is a Pedagogy of Healing Possible?

Understood as an event in the doctor-patient relationship, healing is at first sight what the patient expects from the doctor, but not what he always obtains from him. There is thus a discrepancy between the patient's hope regarding the power that he attributes to the doctor on the grounds of the latter's knowledge and the doctor's recognition of the limits of his own efficacy. There, without doubt, lies the main reason why, of all the objects specific to medical thought, healing is the one that doctors have considered the least. Yet this is also due to the fact that the doctor perceives in healing an element of subjectivity, a reference to the beneficiary's evaluation of the process, when from his objective point of view, healing is the target of a treatment that can be validated only by a statistical survey of its results. And without making a disparaging comparison to those laughable doctors who would make their patients bear responsibility for therapeutic failure, it has to be acknowledged that the absence of healing in one patient or another does not suffice to induce doubt in the doctor's mind concerning the virtue that he attributes to any prescription. Conversely, whoever claims to speak pertinently about an individual being healed should be able to demonstrate that healing, understood as the satisfaction of the patient's expectation, is really the effect of a prescribed and scrupulously applied therapy. However, it is more difficult to carry out such a demonstration today than it has ever been, as a result of the use of the placebo method, the observations of psychosomatic medicine, the attention paid to the intersubjective doctor-patient relationship, and the comparison by some doctors of the power of their own presence to that of a medication. Today, as regards remedies, how one gives them sometimes matters more than what one gives.[1]

In short, we can say that for the sick man, healing is what medicine owes him, while for most doctors, even today, what medicine owes the patient is the best-studied, best-tested, and most-used treatment currently available. Hence the difference between a doctor and a healer. A doctor who does not succeed in healing anyone would not de jure cease to be a doctor—he continues to be licensed by a diploma that sanctions a conventionally accepted knowledge for the purpose of treating patients whose diseases are outlined in medical treatises in terms of symptomatology, etiology, pathogenesis, and therapy. A healer can be one only de facto, for he is judged on the basis not of his "knowledge," but of his successes. The doctor's relation to healing is the inverse of the healer's. The doctor is publicly licensed to claim to cure, while it is the healing that is experienced and avowed by the patient—even if it remains illicit—that vouches for the "gift" of healing in a man whose power is quite often revealed by the experience of others. There is no need to go among the "savages" to learn this. In France, as well, "natural" medicine [*médecine sauvage*] has always prospered just outside the gates of medical schools.

One should thus not be surprised to find that the doctors who first addressed healing as a problem and a subject of interest were, for the most part, either psychoanalysts or those who found in psychoanalysis a chance to question their own practice and its presuppositions—for example, Georg Groddeck, who, in his *Das Buch vom Es* (*The Book of the It*, 1923), was not afraid to equate medicine and charlatanry,[2] or René Allendy in France.[3] While to the traditional medical eye a cure was the effect of a treatment of the cause and functioned to sanction the validity of a diagnosis and the ensuing prescription—hence of the doctor's own worth—to the psychoanalytic eye, a cure became the sign of the patient's rediscovered capacity finally to be done with his own difficulties.[4] A cure was no longer commanded externally; it became an initiative recuperated by the patient, insofar as the illness was no longer treated as an accident, but rather as a failure of conduct, if not as a conduct of failure.[5]

It is known that according to its etymology, "to heal" is to protect, to defend, to arm—quasi-militarily—against aggression or sedition. The image of the organism offered here is the image of a city menaced by an external or internal enemy. To heal [*guérir*] is to guard, to shelter [*garder, garer*]. This was the idea well before certain contemporary physiological concepts, such as aggression, stress, and defense, entered the domain of medicine and its ideologies. The likening of healing to an offensive-defensive reaction is so profound and originary that it has penetrated even the concept of disease, considered as a hostile reaction to some invasion or disorder. This is the reason why, in certain cases, the temporary intention of therapy is to respect the very illness [*mal*] that the sick man expects to see targeted without delay. The need to justify the apparent connivance of therapeutic intention with disease gave rise to certain writings, best known among them the *Traité des maladies qu'il est dangereux de guérir* (*Treatise on Diseases that It Is Dangerous to Cure*, 1757).[6] Jean-Martin Charcot took up this expression in the conclusion of his *agrégation* thesis, "De l'expectation en médecine" ("On the Expectant Method in Medicine," 1857), which saw disease as, despite itself, its own doctor.[7] Alongside a worn-out Hippocratic tradition that was nevertheless latent in many mechanical and chemical disguises from the seventeenth century to the middle of the nineteenth, this thesis configured the representation of the animal organism as an "economy." Animal economy is the set of rules that govern relations between parts in a whole, just as the association of members of a community is governed for its own good by the authority of a domestic or political leader. Organic integrity was a metaphor for social integration before the metaphor was inverted.[8] Thus emerged the general and persistent tendency to conceive the cure as an end to a disturbance and a return to an anterior order, as attested by all the terms with the prefix "re" that serve to describe its process: "restore," "reconstruct," "reestablish," "reconstitute," "recuperate," "recover," and so on. In this sense, to cure implies the reversibility of phenomena whose sequence constituted the disease; it is a variant of the principles of conservation or invariance on which classical mechanics and cosmology are founded.[9] One sees that thus conceived, the possibility of a cure can easily be contested, except in certain patently benign cases such as a head cold or oxyuriasis,[10] because the restitution or reestablishment of the anterior organic state may prove illusory when confirmation rests on functional testing, rather than simply relying on the satisfaction of the man who has stopped calling himself sick.

In the last quarter of the nineteenth century, physiology began to replace the conception of the organism as a compensatory mechanism or a closed economy with a conception of an organism whose functions of self-regulation and adaptation to its milieu are intimately coupled. If homeostasis may at first sight seem comparable to the spontaneous conservation that was celebrated by medicine in the classical age, it cannot nonetheless be held to be isomorphic with it, insofar as for it, an opening to the outside serves as constitutive of properly biological phenomena. Doubtless, pre-physiological medicine did not disregard the surroundings of the organism, the climate, and the seasons. Hence the theory of constitutions.[11] But this theory concerned only common diseases, epidemics that resembled military campaigns. These took time into account—as claimed Thomas Sydenham, for whom diseases followed "the seasons of the year with as much certainty as some birds and plants."[12] The study of circumstances did not aim to learn of what the disease consisted, but to know with what essence of disease one was dealing and what type of therapeutics one should choose. One would thus be mistaken to treat the old theory of epidemic constitutions as anticipating the theory of milieus outlined by Auguste Comte and developed by the positivist doctors of the Société de Biologie, a theory that was contemporary with the constitution of physiology as a science.[13]

The openness of the organism to its milieu, even if it could never be conceived as a simple relation of passive servitude, came progressively to be understood as subordinate to the organism's maintenance of proper constants, expressed in relations where expenditures and gains in energy are controlled by regulatory loops. But the apparent equilibrium or stationary state of such an open system in no way excludes its submission to the second principle of thermodynamics, to the general law of irreversibility and non-return to an anterior state. Henceforth, all the vicissitudes of a healthy or ill organism, or even one considered cured, would be affected by the stigma of degradation. Despite the persistence of the blurred image of Apollo the Thaumaturge within the symbolic of therapy, no doctor can ignore that no cure is ever a return. Indeed, when Freud, in the most debated part of his work, renewed the concept of return, it was only as a return to death, to the inorganic state that preceded life.[14]

As to its object of origin, thermodynamics is the science of the steam engine. It is also, as to the type of society belonging to the scientific institutions from which it emerged, a science that is characteristic of the first industrial societies, societies of predominantly urban populations in which workers' demographic concentration and labor conditions contributed broadly to the development of infectious diseases and in which the hospital imposed itself as the place of a treatment generalized amid anonymity. The discovery of phenomena of microbial or viral contagion and immunity, the invention of antiseptic techniques, serotherapy, and vaccination by Robert Koch, Louis Pasteur, and their students, furnished public hygiene—which until then remained unarmed—with massively effective means. Paradoxically, it was the success of the first methods of cure founded on microbiology that sparked in medical thought the progressive substitution of the social ideal of disease prevention for the personal ideal of healing the ill. There was almost no absurdity in the hope that a population docile to preventive measures could reach such a state of collective health that no individual would find himself in the situation of having to be treated and cured of any known disease. And de facto, in Western societies at the present time, there are almost no cases of smallpox left to treat, because the systematic application of antismallpox vaccination has achieved the result of rendering itself useless. The image of the skillful, attentive doctor from whom singular patients expect their cure has little by little been eclipsed by the image of an agent who, in response to duties that the collectivity declares that it must assume for the good of all, executes the instructions of a state apparatus charged with watching over the right to health to which that each citizen lays claim.

The advances in public hygiene and the development of preventive medicine were supported by the spectacular successes of chemotherapy, itself founded by the research of Paul Ehrlich during the early years of the twentieth century on the artificial imitation of natural processes of immunity. In the history of therapeutics, this is perhaps the most revolutionary invention. Not only did antibiotics provide a way to cure, but by transforming life expectancy, they also transformed the concept of the cure. The statistical calculation of therapeutic performances introduced into the understanding of the cure an objective measure of its reality. Yet this measuring of cures through a statistically calculated duration of survival was inscribed within a tableau that also figured the appearance of new diseases (cardiopathies) and the increased frequency of old diseases (cancers), conditions whose manifestation was a result of the prolongation of the average lifespan. In this way, the fulfillment of the two ambitions of traditional medicine—to cure diseases and to prolong human life—had the indirect effect of confronting today's doctor with patients preyed upon by a new anxiety about the possibility or impossibility of a cure. Cancer took over from tuberculosis. If prolonging the lifespan only confirms the fragility of the organism and the irreversibility of its degeneration, if medicine's history still opens human history to new diseases, then what is healing? A myth?

Though doctors are ordinarily critical of the popular notion of healing, we are not forbidden to attempt to legitimize it. Our language knows "to heal" as an active verb and also as an intransitive verb—as in "to flower" or "to succeed." In popular terms, to heal is to recover a compromised or lost good, namely, health. Despite the social and political implications of this conception, which are due to the recent fact that health is sometimes perceived as a duty to be observed out of respect for sociomedical powers, health has remained the organic state over which the individual estimates himself to be a competent judge. Even if doctors have reasons to consider illusory the definition of health as life lived in the silence of the organs[15]— recalling that such a silence can mask a lesion that has already developed to an irremediable stage—it is still the case that the capacity to be well, that is to say, to comport oneself well in situations that one must face, is a criterion worth conserving.[16] Health is the a priori latent and propulsively lived condition of all chosen or imposed activity. This a priori is what the science of the physiologist can a posteriori decompose into a plurality of constants vis-à-vis which diseases represent a more radical divergence than the average deviation from a norm. However, in substituting the objective analysis of these conditions of possibility for the lived whole of the living subject's power to face them, one substitutes a language for a mode of expression that has been refused the dignity accorded to a language. The doctor is not far from thinking that his science is a well-constructed language, while the patient expresses himself in popular jargon. But since the doctor was first a human being and lived through an age when he was unsure whether he would become God, a table, or a basin,[17] he still carries a few memories of the original block out of which he was sculpted, and in principle, he has retained a few elements of the popular jargon devalorized by his scientist's language. He may therefore consent to understand that his client's request could be aimed at reassuring himself that he will conserve a certain quality in his state of life, or that he can regain an equivalent state, without worrying whether the objective tests of a cure are positive and in agreement. Conversely, a doctor may not understand a patient who, at the end of a treatment that has been prescribed and carried out and that has achieved the disappearance of an infection or a dysfunction, refuses to consider himself over it and does not call himself or behave as someone healed. In sum, from the point of view of medical practice, which is fortified by its scientificity and technology, many patients are satisfied with less than what one may think they are owed and certain others refuse to recognize that what they were due has been done. Thus, health and healing arise from a genre of discourse other than the one whose vocabulary and syntax we learn in medical treatises and clinical lectures.

When in 1865 Jean Antoine Villemin presented what he considered to be solid proofs that tuberculosis is contagious, he needed to provide much more to obtain the agreement of his contemporaries; for many of them, such as Isidore Bricheteau, who alluded to the draconian ordinances that had been in place since the nineteenth century in Spain and the Kingdom of the Two Sicilies, the idea of contagiousness could have been born only in the Southern imagination.[18] Doctors had somehow managed to integrate a popular reaction of dread and rejection into their conception of disease, even while they were struggling against it. This is because in human tuberculosis and bovine or avian tuberculosis, whose identity or difference was still being debated, medicine established the active presence of a determinant that, for lack of a better term, should be called "psychological."[19] Tuberculosis was a source of terror, much as leprosy had been in the Middle Ages. Naming the disease would aggravate its symptoms, because the disease entailed social exclusion as much as organic consumption.[20] For a long time, one would become ill from being cured of such a disease, in the sense that one would perceive all around oneself a suspicion of residual noxiousness. Even when monitored by laboratory tests, healing was not accomplished through reintegration into social existence, more because of the anxiety of segregation than because of the reduction of vital capacities. This form of healing, which could be called "pathological," is rarer today in tubercular cases, but has become frequent in cancer cases because of a similar anxious reaction to the idea that the cured person's circle is supposed to have of this unforgiving disease. Meanwhile, in addition to patients who simply do not manage to assume their own cure, who do not comport themselves as cured and resolved to confront anew, however differently from before, the question of existence, there are patients who find some good in their disease and refuse the cure. In this passive resistance to medical intervention, the patient seeks a kind of compensation for his diminished, dominated condition. In the therapeutic relation, he assures himself of keeping the initiative.[21]

This rehearsal of pathological configurations in which it is not possible to envisage a cure in the traditional sense of an end and a new beginning forbids us to conceive the doctor-patient relation as the relation of a competent technician to a machine that is out of order [*dérangé*]. And yet, the formation of doctors in medical schools prepares them very poorly to admit that healing is not exclusively determined by interventions of a physical or physiological order. No illusion of professional subjectivity is worse for doctors than their confidence in the strictly objective foundation of their counsels and their therapeutic gestures—their contempt for or their self-justifying forgetfulness of the active relation (whether positive or negative) that cannot fail to arise between the doctor and the patient. During medicine's positivist age, this relation was considered an archaic residue of magic or fetishism. The reactualization of this relation should be credited to psychoanalysis, something studied too many times for it to be useful for us to return to it here.[22] But it still seems urgent to ask what place can be claimed by the attention that any particular doctor accords to any particular sick man in a medical space increasingly occupied, in the so-called developed nations, by sanitary equipment and regulations and by the programmed multiplication of "curing machines" [*machines à guérir*].[23]

Things had reached the point that my brain could no longer bear the worries and torments that were inflicted on it. It said: "I give up; but if anyone here cares about my preservation, let him relieve me of a tiny bit of my burden and we will make some more time." At this moment, when the brain apparently has little left to lose, the lung presents itself. These debates between the brain and the lung, which took place unbeknownst to me, must have been frightful to witness.[24]

And:

my attitude toward the tuberculosis today resembles that of a child clinging to
the pleats of its mother's skirt. . . . I am constantly seeking an explanation for
this disease, for I did not seek it. Sometimes it seems to me that my brain and
lungs came to an agreement without my knowledge.[25]

Not all patients, and not all tuberculars in particular, are Kafkas. But
who would not recognize, in the confidences of the author of *The Trial*, the
truth of these situations of distress, which have a psychosocial origin and
generate the type of organic exhaustion conducive of the outbreak of infectious disease? Even more so if we are speaking of afflictions that affect the
neuroendocrinal system, from chronic fatigue to gastrointestinal ulcers and
more generally the so-called adaptation disorders.

Because these situations of distress are often manifestations of the block-
ing of social structures of communication, wouldn't the study of potential
remedies for them arise only from disciplines of a sociological order? What,
then, would be the type of society that, supplied with a health organization
that exploits the most sophisticated information about the distribution and
correlations between factors of disease, would one day relieve the doctor of
the perhaps desperate task of helping individuals in situations of distress in
their anxious struggle for aleatory healing?

And why, finally, should we really try so hard to hide from people that it
is normal to fall ill from the moment one is alive, that it is normal to heal,
with or without the help of medicine, that disease and healing are inscribed
within the limits and powers of biological regulation? Biological normalities
have no guarantee other than their fact, unless one gives them a metaphysical foundation, which nothing forbids us from seeing as only a consecration
of that fact. Life must be a given for us to believe in the possibility that it is
a necessity.

The organism of a living being may undergo alterations of structure or
disturbances of functions that, if they do not go so far as to destroy it, can
still compromise its execution of tasks imposed by its specific heredity. But
the task specific to man has been shown to be the invention and renewal of
tasks whose exercise requires both apprenticeship and initiative within a
milieu that is itself modified by the very results of this exercise. Man's diseases are not only limitations of his physical power, they are the dramas of
his history. Human life is an existence—a being-there for a becoming that

is not preordained and is haunted by its end. Thus, man is open to disease not by being condemned to it or destined for it, but because of his simple presence in the world. In this regard, health is not at all an economic exigency to be asserted within a legislative framework; it is a spontaneous unity of the conditions for the exercise of life. This exercise, on which all other exercises are founded, founds for them and restricts, as they also do, the risk of failure, a risk from which the individual cannot be protected by any statute of socially normalized life. Health insurance, invented and institutionalized by industrial societies, finds its justification in the project of procuring for man—who now becomes assured of being compensated for his potential economic losses—a confidence and audacity in accepting tasks that, to some degree, always carry a risk to life. It is appropriate, then, to work today to cure men of the fear of eventually striving, without guarantee of success, to heal from diseases whose risk is inherent in the enjoyment of health.[26]

On this point, it is perhaps surprising that Kurt Goldstein's thesis, developed in *The Organism* (1934), has echoed so little outside of philosophical circles influenced by Maurice Merleau-Ponty's works.[27] Perhaps this is because Goldstein himself presented his thesis as an epistemology of biology, rather than as a philosophy of therapeutics. Nevertheless, in the final pages of his work, the doctor's activity is likened to that of the pedagogue.[28] Goldstein developed concepts of ordered behavior and catastrophic behavior on the basis of observations concerning the behavior of men suffering from cerebral lesions.[29] A healthy organism comes together with the surrounding world in such a way as to be able to realize all of its capacities. The pathological state is the reduction of the initial latitude for intervening in the milieu. The anxious effort to avoid situations generative of catastrophic behavior, the tendency toward the simple conservation of a residue of power, is the expression of a life that is losing its "responsiveness." If by "healing" we understand the set of processes through which the organism tends to overcome the limited capacities to which disease constrained it, then we should admit that to heal is to pay, in effort, the price for slowing the process of degradation. "Thus, the patient frequently has the choice whether he wants to accept—corresponding to the change caused by the disease—a limitation of the milieu and the resulting limitation of freedom, or less limitation and more suffering instead. If the patient bears more suffering, he will gain in possibilities of performing since therapeutic measures may be apt to reduce suffering but at the same time diminish perform-ances."[30] Under these conditions, what stance should the doctor adopt—that of a counselor, or of a guide? Goldstein thus heralds the questions to which Michael Balint's work has brought a perhaps less well-founded notoriety.[31] The doctor who decides to guide the patient along the difficult path of healing

> will be able to do so only if he is completely under the conviction that the doctor-patient relationship is not a situation depending alone on the knowledge of the law of causality but that it is a coming to terms of two persons, in which the one wants to help the other gain a pattern that corresponds, as much as possible, to his nature. This emphasis on the personal relationship between doctor and patient marks off, impressively, the contrast between the modern medical point of view and the mere natural-science mentality of the physicians at the turn of the century.[32]

Rather than be surprised, which is facile, one should seek to comprehend. The indifference or hostility of the great majority of doctors to the questions posed by certain movements of dissent within their profession—questions concerning their abandonment of their healing vocation in favor of the regulated tasks of screening, treatment, and control—could be explained in the following manner. Nothing is more widespread and profitable today than *anti-x* proclamations. It was antipsychiatry that started the trend; antimedicalization followed. Well before Ivan Illich's exhortations for individuals to reclaim the regulation of their health, to self-manage their own healing, and to reclaim their own death, the reduction of psychoanalysis and psychosomatics to the level of media vulgarization popularized the idea that converting the patient into his own doctor is desirable and possible.[33] One thought one was inventing something while merely reprising the thousand-year-old theme of being one's own doctor.[34] Because the times are hard and prospects are few, a rising number of practitioners of nonscientific therapies (voilà the enemy—science!) flatter themselves for having achieved what they reproach doctors for neglecting or missing. Hence this appeal to disappointed patients: come and tell us that you want to heal, and together, we will do the rest. The arguments are sometimes so hollow, so vainly peremptory, that one could almost regret the progressive disappearance of the species of doctors who, as Goldstein put it, had habits of thought proper only to the physical sciences. And we can see now why the conceptual triviality of the propagandists of self-cures pushes many doctors, already ill at ease in the role of often-powerless therapist, away from an ideology that is so well-intentioned, yet has so little concern with self-critique.

Like antipsychiatry, antimedicine exploits the initial advantage gained from arguments that assume as true what they need to prove.[35] Let us suppose the problem is resolved: let us make Brutus Caesar! Perhaps Brutus suffers from lingering pains in the region of the stomach that, for long periods, recur violently every day.[36] Medical disinformation has taught him about the symptoms of an ulcer, about the effects of the emotions on hormonal secretions. He has heard about the epidemic of gastric ulcers among the population of London during the bombings of the last world war. Will Brutus first tell a psychotherapist about his marital difficulties with Portia, or will he seek out the nearest radiologist? While deciding between the two, and in order to ease the pain, will he adopt a strict diet and take bismuth salts? As we can see, Brutus has become, unbeknown to himself, the mirror on which the faces of different doctors are reflected and blur away. He who claims to be liberated from technocracy finds himself trapped by the threads of a medicine that continues to search how to weave them best. Brutus can escape by going to a healer.

In sum: if doctors, who are already concerned with keeping their competence up to date, neglect (less perhaps on principle than for lack of time) to inquire patiently about the potential affective distress of their clients, must one conclude that they are inferior to the first therapist to come along and appeal to psychosomatics? Would the latter be better qualified to treat a case of obesity first brought on by eating habits caused by affective compensation, but now determined by a thyroid or adrenal disorder? Is psychologism any better than physiologism in matters of therapeutic reductionism?

Let us then treat as resolved the problem of the time needed for long therapeutic interviews, which comes back to the problem of the inevitable multiplication and the remuneration of doctors educated to listen to the embarrassed complaints of their clients. Should one introduce into the university-hospital education of future doctors instruction in "convivial" participation and thus tests and exams in aptitude for human contact? Or along a different path, should one resolve the difficulty by creating health-care teams in which some highly motivated doctors and paramedical personnel apply themselves to re-creating the relation of individuals to their bodies, to their work, and to the collective? Are these solutions, which claim to belong to the Left, exempt from all collusion with an ideology of the Right? Human contact is neither taught nor learned in the same way as the physiology of the autonomic nervous system. To keep away from the medical profession whoever is not gifted in "convivial" participation would amount to instituting a new criterion of inegalitarian selection. In a team of health-care workers, one finds people who have the responsibilities of engineers and others who are content to be supervisors. Finally, is any systematic campaign to demedicalize health assured that it will not obtain the opposite result? In promising a better individual use of better collective conditions of health, modeled after a more equitable redistribution of wealth, can one be sure that one is not giving rise to an obsessional neurosis with health? To consider oneself deprived by the current state of medical practice of the health that one deserves is itself a kind of illness.

It is one thing to obtain the health that one believes one deserves; it is another to deserve the health that one obtains for oneself. In this latter sense, once the doctor has prescribed the treatment required by the organic state of the patient, the part that he can play in healing would consist in instructing the sick man in the patient's responsibility to achieve a new state of equilibrium with the demands of the environment. The doctor's objective, like that of the educator, is to render himself useless.

The indiscriminate celebration of the virtues of natural medicine [*médecine sauvage*] does not seem indispensable to the purpose of critiquing certain practices of the civilized medical profession. But it seems that the time has come for a *Critique of Practical Medical Reason* that would explicitly recognize, within the ordeal of healing, the necessity of collaboration between experimental knowledge and the propulsive nonknowledge of this a priori opposition to the law of degradation for which "health" expresses a success that always becomes suspect again. This is why, if a pedagogy of healing were possible, it would have to include the equivalent of what Freud called "reality testing."[37] This pedagogy should seek to obtain the subject's recognition of the fact that no technique, no institution, whether currently or in the future, will assure the integrity of his powers of relating to men and things. The life of the individual is, from its beginnings, the reduction of life's powers. Because health is not a constant of satisfaction, but the a priori of the power to master perilous situations, this power uses itself up in mastering successive perils. Health after healing is not the same health as before. The lucid consciousness of the fact that healing is not a return helps the patient in his search for the state of the least possible renunciation by liberating him from his fixation upon his previous state.

One of F. Scott Fitzgerald's last texts, "The Crack-Up," begins with these words: "Of course all life is a process of breaking down." A few lines later, the author adds: "The test of a first-rate intelligence is the ability to hold two opposed ideas in the mind at the same time and still retain the ability to function. One should, for example, be able to see that things are hopeless and yet be determined to make them otherwise."[38]

To learn to heal is to learn the contradiction between today's hope and the defeat that comes in the end—without saying no to today's hope. Is this intelligence, or simplicity?

The Problem of Regulation in the Organism and in Society

When my friend Pierre-Maxime Schuhl[1] asked me to lecture at a meeting of the Alliance Israélite Universelle, I accepted gladly and with great pleasure; it is an honor for me, and I only regret having had to pose this one condition—for which I apologize—which resulted in us meeting at a time so out of the ordinary.

I have chosen to treat a problem that, I assure you gentlemen, I have by no means mastered, for it is a question for me, as well. But I have chosen to speak to you of a subject that is not worrying because it worries me, but worries me because I believe it to be fundamentally worrisome. All told, behind the rather too technical title "The Problem of Regulation in the Organism and in Society" is to be found nothing less than a very old, still-unresolved problem, that of the relations between the life of the organism and the life of a society. Is the frequent comparison [*assimilation*] of society to an organism—sometimes scholarly, sometimes popular—anything more than a metaphor? Does this comparison overlie a substantial kinship?

Naturally, this problem is of interest only to the extent that the solution given to it becomes (if it is positive) the point of departure for a political theory and a sociological theory that tends to subordinate the social to the biological and to the extent that it in fact becomes—I will not say risks—an argument for political practice. That there, as a result, we find a subject of considerable concern—this I do not think I need to declare and demonstrate more fully.

This permanent comparison of society to an organism derives from a temptation that is, in general, doubled by the inverse temptation—that of comparing the organism to society.

In the early stages of biological philosophy, one of the Greek thinkers in whom Schuhl has been interested, Alcmaeon of Croton, interpreted the disequilibrium caused by disease, by pathological disorder, as sedition. That is, to explain the nature of the disease in the organism, he brought over a concept of sociological and political origin.[2]

When liberal and socialist economists in the eighteenth and nineteenth centuries drew attention to the social phenomenon of the division of labor and its effects—fortunate, according to the former, detestable, according to the latter—physiologists found it natural to speak of the division of labor concerning the cells, the organs, and the devices [*appareils*] that make up a living body.

At the moment of cell theory's diffusion in the second half of the nineteenth century, Claude Bernard spoke of the "social life" of cells; he asked himself whether cells have the same life in society as they would have in freedom, which came back to anticipating the problem posed by the results of a cell culture.[3] Does the cell, when freed from all the relations that it maintains with others in an organism, behave the same way in freedom as it would in society?

Ernst Haeckel, one of those who did the most to elevate cell theory to the rank of a dogma, spoke of a "Cell State" and a "Republic of Cells" to designate the body of the multicellular living being.[4]

In short, in the movement from sociology to biology, it is unnecessary to multiply examples further in order to reinforce this idea.

Here, we must remark that there has always been an exchange of ambiguous figures of speech between sociology and biology. In some cases, only history allows us to clarify the origin of certain concepts for which a certain equivocality in biology and in sociology gives an impression of equivalent validity in each of these domains of signification and use.

For example, there is a concept that is fundamental to politics and economics—the concept of *crisis*. Well, it is a concept of medical origin—it is the concept of a change that, signaled by certain symptoms, intervenes in the course of an illness and that will indeed decide the life of the patient.

I remind you also that the term "constitution," which is also one of these perfectly equivocal terms, is as valid on the biological as on the social field; if we look for a passage from one field to another, from the biological to the social, we will not find it, however far back we may go. The term has always been ambiguous, equivocal—it works as well for one field of explanation as for the other.

I recall all these facts only to show that when one compares society to an organism, it is not only as a result of a short-lived sociological theory at the end of the nineteenth century, a theory whose days were rapidly numbered. This theory is called organicism. That this theory would appear in explicit form only at that particular moment had by no means prevented certain sociologists, such as Auguste Comte, from seeking in the concept of "consensus," of sympathy between the parts of an organism—a concept that is of biological origin—a notion that he could import into the sociological field, even as he recognized that, because of human history, because of the fact of tradition, social life and organic life compose two radically heterogeneous domains.

That said, we approach the problem via what I could call its most popular aspect, that is, the temptation of mutual comparison, and I would like to show that if we indeed place ourselves at the point of view of popular representation, a correction for this comparison immediately stands out. By this, I mean that insofar as the social problem and the problems posed by organic life and its disorder are concerned, there is, in public opinion, an attitude that should by itself invite the philosopher to inquire about its underlying reasoning.

Of course, the problem in comparing society to an organism has interest only to the extent that through it we understand certain views concerning the structure or functioning of a society, but even more concerning the reforms to be carried out once the society in question is affected by grave disorders; differently put, what dominates the comparison of the organism to a society is the idea of social medication, the idea of social therapeutics, the idea of remedies for social ills.

We should then remark that in the link of heath to disease and thus in the relation [*rapport*] to the repair of organic or social disorders, the relations between the illness and the remedy are radically different when an organism is concerned and when a society is concerned.

There is nothing mysterious in what I'm alluding to—everyone has experienced it, if I may say so; it fuels everyday conversations. An organism is a mode of being that is exceptional in that there is, strictly speaking, no difference between its existence and its ideal, between its existence and its rule or norm. From the moment that an organism is, from the moment that it lives, it is what is possible, which is to say, it responds to an organismic ideal [*idéal d'organisme*]; the norm or rule of its existence is given in its existence itself, in such a way that, when what is in question is a living organism, and to take the most banal example, when what is in question is the human organism, the norm one must restore when this organism is harmed or ill does not in the least lend itself to ambiguity. We know very well what the ideal of a sick organism is; the ideal of a sick organism is a healthy organism of the same species. Which is to say that even though we may not know exactly of what an organic disorder consists, even though the physician may debate the nature of the ill [*mal*], and even though one may debate the composition and administration of remedies, no one debates the effect expected of these remedies—the restoration of the organism to its healthy state. In brief, the ideal of the organism here is clear to everyone—it is the organism itself. One may hesitate about the diagnostics and the therapeutics of an affliction of the liver or a disease of the eyes, but no one hesitates about what we should expect from therapeutics; we expect the liver to secrete bile and the eyes to have a satisfactory acuity. In short, in the order of the organism, we commonly see the whole world debate the nature of evil [*mal*] and no one debate the ideal of the good.

But the existence of societies, of their disorders and unrests, brings forth a wholly different relation between ills and reforms, because for society, what we debate is how to know its ideal state or norm.

It precisely here that the problem is posed; the purpose [*finalité*] of the organism is interior to the organism, and in consequence, the ideal we must restore is the organism itself. For its part, the purpose [*finalité*] of society is precisely one of the capital problems of human existence and one of the fundamental problems that reason poses to itself. Ever since man has lived in society, it is precisely the ideal of society that all the world debates; by contrast, men agree much more easily on the nature of social ills than on the scope of the remedies to apply to them. In the existence of a society, the norm of human sociability is not determined [*enfermée*], and I will try very soon to say why. Hence the multiplicity of possible solutions calculated or dreamt by men to put an end to injustices. One could say that in the organic order, the use of an organ, a device, an organism is patent; what is sometimes obscure, what is often obscure, is the nature of a disorder. From the social point of view, it seems on the contrary that abuse, disorder, and evil [*mal*] are clearer than normal circumstances. Collective agreement is more easily reached on disorders: child labor, bureaucratic inertia, alcoholism, prostitution, police arbitrariness—these are social ills on which collective attention falls (that is, among men of good faith and good will) and on which collective agreement is easy. On the contrary, the same men who agree on a social ill part ways on the subject of reforms, so what appears to some as a remedy appears to others as a state worse than the ill itself, precisely because of the fact that the life of a society does not inhere in society itself.

One could say that in the social order, madness is more clearly discerned than reason, whereas in the organic order, it is health that is more easily discerned and determined than the nature of the illness. This idea has been the object of brilliant (a bit too brilliant) developments on the part of an English author, Gilbert Keith Chesterton, in a work that is not well known, but that has been translated into French, *What's Wrong with the World* (1910).[5] On this subject, Chesterton contented himself, as was his habit, with pointing out very exciting and very stimulating paradoxes. But describing is not enough. I do not say that I will explain—I do not pretend to do so—but I would like to show how, starting from an observation accessible to every man of good will, one may found certain principles of explication.

This is where the word "regulation," which appears in the title of my lecture, intervenes; it is a scientist's word, though not really, insofar as everyone knows what a regulator in an old locomotive is and everyone knows what a regulating station [*gare régulatrice*] is.[6] The concept of regulation is a concept that I would not call familiar, yet not forbidding, either.

The living organism is a type of being that is characterized by the constant presence and permanent influence of all its parts on each of them.

What is proper to an organism is to live as a whole and not to be able to live except as a whole. This is made possible by the existence in the organism of a set of apparatuses or mechanisms of regulation whose effect consists precisely in the maintenance of this integrity, in the persistence of the organism as a whole. This idea of organic regulation is a rather recent concept; I will soon offer some examples of the principal types of organic regulation.

This idea, which begins with the physiology of Claude Bernard, only confirms a very old intuition of Hippocratic medicine—that there exists, by the very fact of the organism's life, a sort of natural medication or natural compensation for the lesions or the disorders to which the organism may be exposed. Modern physiology has provided only confirmations to this old Hippocratic idea of the healing power of nature. An organism comprises, by the sole fact that it is an organism, a system of mechanisms of correction and compensation for the divergences and injuries to which it is subjected by the world in which it lives—by its milieu, a milieu vis-à-vis which these mechanisms of regulation allow the organism to lead a relatively independent existence. To take a very simple example, I cite what used to be called "cold-blooded" and "warm-blooded" animals, which today are called, in a more scientific fashion, "poikilotherms" and "homeotherms." In cold-blooded animals, there is no system to regulate temperature; they are slaves of the temperature of the milieu. The homeotherm has a system of regulation that allows it to compensate for differences between its temperature and that of the milieu, to maintain a constant temperature independent of the milieu's prompts.

By the sole fact of its existence, the organism resolves on its own a kind of contradiction, the contradiction between stability and modification. The expression of this original fact requires terms whose signification is at once physiological and moral; there is in every organism an inborn moderation, an inborn control, an inborn equilibrium; it is the existence of this moderation, this control, this equilibrium, that, following the American physiologist Walter B. Cannon, is known by the scientific term "homeostasis."[7]

The organism's stable states are obtained in all its parts by conserving the uniformity of the natural conditions of the life of these parts—that is, by preserving it from divergences too great from within or from without. In other words, the organism's stable states are obtained by conserving what, since Claude Bernard, is called the "internal milieu." Just as the notion of milieu served biologists at the end of the eighteenth century and the beginning of the nineteenth to explain the modifications and adaptations of the organism and of species, this notion of internal milieu allowed Claude Bernard to explain how, in the organism's interior, each part finds itself in relation to all the others by the intermediary of a sort of liquid matrix composed of salts, water, and products of internal secretion, a matrix whose stability depends on two devices that in the higher animals are the keystones of all these operations: the nervous system and the system of internal secretion (or endocrine) glands. Bernard's originality resided not just in his showing that there exists an internal milieu, but indeed in his showing that it is the organism that produces this internal milieu. I insist here on the fact that the regulation of the organism is ensured by the special devices that are the nervous and endocrine systems. The regulations that interested Bernard were physiological regulations, for example, the regulation of respiratory movements by the level of carbonic acid contained in the internal milieu; the regulation of the removal of water and salts that annuls the variation of osmotic pressure in internal fluids; thermoregulation (the regulation of animal heat); or the regulation of divergences of nitrogenous nourishment by the maintenance of the law of nitrogen equilibrium.[8]

Claude Bernard's research in two further areas—embryonic development and regeneration—compounded the studies we have been discussing.

Embryologists have discovered that in the fertilized egg, in the course of embryonic life beginning with this fertilized egg, there exists a sort of control by a totality over its parts, thanks to which whatever the variations of ovular substance (if I may put it this way), the living being conserves or maintains the integrity of a specific form. For example one may, using half of an egg or (on the contrary) two joined eggs, obtain a single individual whose specific characteristics are identical to those one could obtain through the development of a normal egg, with only some quantitative differences.[9]

Here, the regulation of what has been called the "specific organizers" functions in such a way that by comparison with the injuries to which the egg is subjected to by exterior elements, the specific form to be obtained is constantly preserved and maintained.

In the same way (and this is only a consequence), the regeneration that occurs in some animals and that enables these animals to recover their proper form following a mutilation, with only some quantitative differences, shows well that there is a sort of domination of form over matter, a sort of command of the whole over the parts.

All of this is to say that it is not without profundity that a biologist that I just mentioned, Cannon, could title the work in which he summarily outlines these mechanisms of regulation *The Wisdom of the Body*.[10] It is a title we can sneer at, but that is nevertheless worth reflecting on.

What was, in effect, the ancient and pagan idea of wisdom?

In order not to take a lashing from my friend Schuhl's critiques, I will pass over this question rapidly, and I will say that the idea of wisdom was essentially an idea of the measure of control and of mastery in the conduct of life. This is what protected man from the thrall of immoderation—the permanent temptation of deviance, aberration, and contempt for the limit.

It is certain that for many Greek thinkers—some of the greatest ones included—the idea that they had of the universe, the idea they had of the Whole, was the idea of a healthy organism, which is to say an organism all of whose parts accord with one another, are present to one another, and maintain invariable functional relations with one another. Inside this Whole—inside this order that is at the same time life—every being, man included, has a place; in this place, it must work in cooperation with the ensemble of other beings; it must always respect the functional relations that subjugate it to the exigencies of the Whole.

This idea of ancient wisdom is perhaps an idea grafted on an image borrowed from the intuition of life. Of course, it is not the body that is wise, but reason. But as soon as we speak of the wisdom of the body, we restore to the body the image of equilibrium. This is the image on which the idea of wisdom is developed, and, I would say, on which it is grafted.

Cannon's book includes an epilogue entitled "Relations between Biological and Social Homeostasis." Cannon here gives in to the penchant proper to every specialist; he gives in to the temptation that the scientist shares with the common man, which is to import into sociology this magnificent concept of regulation and homeostasis, whose mechanism he has described in the course of the preceding pages.

This book by Cannon, I will say immediately, because this connection is not uninteresting, is a collection of lectures he gave at the Sorbonne in 1930. He was at the time a professor at Harvard. Now, 1930 was the year that Bergson must have been putting the last touches on (perhaps correcting the proofs) of the *Two Sources of Morality and Religion*.[11] We are thus almost certain that there was no influence. But Cannon's works antedate this considerably, and Bergson, who read everything and knew of everything, may have known of them. What is interesting is to see that in the years from 1930 to 1932, Cannon and Bergson encounter the same problem, the first starting from his biology, the second starting from his philosophy.

It really must be said that Cannon's epilogue on social homeostasis is the weakest part of his book. First of all, it's the shortest; we could say that he was being modest, that he was outside his domain and thus proceeded carefully, but allowing that it is the shortest, it is still the weakest, because the majority of comparisons are founded on commonplaces of politics and sociology, whose foundation he does not seek.

Cannon asks himself if one could not find in society examples of mechanisms of regulation that absorb divergences and tend to compensate for disorders.

Here is one example that I will take the liberty to read in full:

> At the outset, it is noteworthy that the body politic itself exhibits some indications of crude automatic stabilizing processes. In the previous chapter I expressed the postulate that a certain degree of constancy in a complex system is itself evidence that agencies are acting or are ready to act to maintain that constancy. And moreover, that when a system remains steady it does so because any tendency towards change is met by increasing effectiveness of the factor or factors which resist the change. Many familiar facts prove that these statements are to some degree true for society even in its present unstabilized condition. A display of conservatism excites a radical revolt and that in turn is followed by a return to conservatism. Loose government and its consequences bring the reformers into power, but their tight reins soon provoke restiveness and the desire for release. The noble enthusiasms and sacrifices of war are succeeded by moral apathy and orgies of self-indulgence.[12]

And now we come to a passage to which I ask you pay some attention: "Hardly any strong tendency in a nation continues to the stage of disaster; before that extreme is reached corrective forces arise which check the tendency and they commonly prevail to such an excessive degree as themselves to cause a reaction."[13] I cannot help comparing this remark by Cannon to much more profound remarks that Bergson makes, at the end of *Two Sources of Morality and Religion*, on what he calls the "law of dichotomy" and the "law of double frenzy."[14]

For Bergson as well, society, which as you know is at once closed and open, that is to say, conservative, tending to its own conservation like an organism, yet searching ultimately to overcome itself toward Humanity, like the impetus [*élan*] that carries universal existence across matter in an infinite current of creation—society is at each moment in its history oriented by some tendency or other; one tendency carries it toward another, yet once it attains a kind of climax, it is the opposite tendency that in its turn deploys itself.

Still, Bergson does not reason like Cannon, who seems to hold, in the wisdom of the social body, to a kind of extension of Le Chatelier's Principle, which claims that when certain perturbations tend to exert themselves in a system of movement, resistance to these perturbations is produced by relations internal to the system. Bergson, on the contrary, says that if in a certain sense there exists an oscillation around a median position, a sort of pendular movement, then the pendulum, insofar as society as concerned, is endowed with memory, and the phenomenon is no longer the same on the return as it was on the way.[15] Moreover, we should say that in the example invoked by Cannon, the conservative and reformist alternation does not have meaning in every society; but only in a parliamentary regime, that is to say, in a political apparatus that is a historical invention made to channel discontent.[16] It is a type of apparatus that is not inherent in social life as such; it is a historical acquisition, a tool that a certain society gave itself.

Having pronounced this word "tool," I will rapidly set out the reasons why we cannot conceive society to be like an organism.

Concerning society, we must address a confusion that consists in the confounding of organization and organism. That fact that a society is organized—and there's no society without a minimum of organization—does not mean that it is organic; I would gladly say that organization at the level of society is of the order not of organic organization, but of design. What defines the organism is precisely that its purpose, in the form of its totality, is present to it and to all its parts. I apologize—I will perhaps scandalize you—but a society has no proper purpose; a society is a means; a society is more on the order of a machine or of a tool than on the order of an organism.

Certainly, a society bears some resemblance to what is organic, since it is a collectivity of living beings. We cannot, properly speaking, decompose a society, but if we analyze it, which is a very different thing, we discover that while a society is a collectivity of living beings, this collectivity is neither an individual nor a species. It is not an individual, because it is not an organism endowed with a purpose and a totality that are obtained by a specialized system of devices of regulation; it is not a species, because it is, as Bergson says, closed. Human societies are not the human species. Bergson shows that the human species is in search of its own specific sociability. Thus, society, being neither an individual nor a species, but a being of ambiguous genus, is as much a machine as it is a living thing; not being its own end, it simply represents a means, it is a tool. Consequently, not being an organism, society presupposes and even calls for regulations; there is no society without regulation, and there is no society without rules, yet in society, there is no self-regulation. There, regulation is always, if I may say so, something added on and always precarious.

One could ask in this case, and without paradox, whether the normal state of a society is disorder and crisis, rather than order and harmony. In saying "the normal state of society," I wish to say the state of society understood as a machine, the state of society understood as a tool. It is a tool that is always out of order, because it is deprived of its specific apparatus of self-regulation. In saying "the normal state," I did not wish to suggest it to be the ideal of human life. The ideal of human life is neither disorder nor crisis. But this is precisely why the supreme regulation of social life, namely, justice, does not figure in the form of an apparatus produced by society itself, even if there exist in society institutions of justice.

In society, justice has to come from elsewhere, and this is what Bergson has shown. The Bergsonian idea is much more profound than it appears, even to a serious and attentive reading (I will not say to a quick reading, for that does not lead us to understanding). I wonder whether the distinction and the opposition that he establishes between wisdom and heroism does not dovetail with this idea that justice cannot be a social institution, that it is not a regulation inherent in society, but a different thing altogether. Already in Plato, justice was not inherent in one part of the social body; it was the form of the whole. If justice, the supreme form of the regulation of human society, is not inborn in society itself, then it is not exercised by an institution that is on the same level as the other institutions. This perhaps helps us to understand one fact: there is no social wisdom in the way there is an organic wisdom. One does not need to become clairvoyant because one is born in a certain species that has eyes, that cannot move and cannot live unless it moves in the light (whereas a plant that lives merely by growing in the light). From the moment that one has eyes, one sees; yet one is not wise in the way that one sees with one's eyes; there is no social wisdom in the way there is a wisdom of the body. One must become wise, one must become just. The objective sign that there is no spontaneous social justice, that is to say, no social self-regulation, that society is not an organism and as a result that its normal state is perhaps disorder and crisis, this is the need for a hero that societies periodically feel.

Wisdom and heroism are strangers to each other. Where there is wisdom, we have no need for heroism, and where heroism appears, it is because wisdom has been missing. Differently put, it is the absence of social wisdom, the absence of social homeostasis, the absence of those regulations that make an organism be an organism—it is precisely the absence of all this that explains the way in which the arrival of a social crisis can suggest that the very existence of society is threatened. At this moment we find what Bergson calls "the call of the hero."[17] Wise men not having resolved the problem, not having avoided its being posed, the hero is he who will find or invent a solution. Naturally, he can invent it only in extremity, can invent it only in the midst of peril.

This is the reason why I believe that there is an essential link between the idea that justice is not a social apparatus and the idea that up to the present, no society has managed to survive except through crises and thanks to these exceptional beings who are called heroes.

Under these conditions, if I have not proved to you, if I have not accomplished—and I'm quite far from doing so—this tour de force to which your president too generously alluded, if I have not succeeded in proving to you that society is not an organism (and besides, in these matters there is no proof), that one must not allow it to resemble an organism, that we must thus be vigilant toward all these comparisons [*assimilations*] whose consequences you can guess—if I have not succeeded in demonstrating this to you, I would simply be happy if I have at least managed to pose for you certain problems, the same ones I pose for myself, in a form that would appear to you worthy of reflection.

INTRODUCTION: GEORGES CANGUILHEM'S CRITIQUE OF MEDICAL REASON, BY
STEFANOS GEROULANOS AND TODD MEYERS

1. To give but one, perhaps unexpected example: despite its difference in
both argument and genre, François Jacob's classic history of biology and heredity, *The Logic of Life*, shows clear signs of Canguilhem's influence, for instance
in its discussion of classical biological mechanism. See François Jacob, *The Logic
of Life: A History of Heredity*, trans. Betty E. Spillmann (New York: Pantheon,
1973).

2. Georges Canguilhem, *Essai sur quelques problèmes concernant le normal et le
pathologique* (Clermont-Ferrand: La Montagne, 1943) and *Le normal et le pathologique, augmenté de Nouvelles réflexions concernant le normal et le pathologique*
(Paris: Presses Universitaires de France, 1966); available in English as *The Normal and the Pathological*, with an introduction by Michel Foucault, trans. Carolyn
R. Fawcett in collaboration with Robert S. Cohen (New York: Zone Books,
1991). Subsequent references to this text are to the English translation.

3. Georges Canguilhem, *La formation du concept de réflexe aux XVIIe et XVIIIe
siècles* (Paris: Presses Universitaires de France, 1955).

4. Georges Canguilhem, *La connaissance de la vie* (Paris, Hachette, 1952) and
2nd, augmented edition (Paris: Vrin, 1965), available in English as *Knowledge of
Life* (New York: Fordham University Press, 2008). Subsequent references to
this text are to the English translation.

5. Georges Canguilhem, *Idéologie et rationalité dans l'histoire des sciences de la
vie* (Paris: Vrin, 1977), available in English as *Ideology and Rationality in the History of the Life Sciences*, trans. Arthur Goldhammer (Cambridge, MA: MIT Press,
1988). Subsequent references to this text are to the English edition.

6. Georges Canguilhem, *Études d'histoire et de philosophie des sciences*, 5th ed.
(Paris: Vrin, 1983).

7. See the overviews of Goldstein's interwar work in Anne Harrington, *Reenchanted Science: Holism in German Culture from Wilhelm II to Hitler* (Princeton: Princeton University Press, 1999) and Mitchell Ash, *Gestalt Psychology in German Culture, 1890–1967: Holism and the Quest for Objectivity* (Cambridge: Cambridge University Press, 1995). Canguilhem would cite Goldstein on this argument from his earliest work in *The Normal and the Pathological* until his last essays.

8. Canguilhem, *The Normal and the Pathological*, 185 and 188.

9. Ibid., 91.

10. Canguilhem, "The Normal and the Pathological," in *Knowledge of Life*, 126 and 132. See also the "absolute normative" and "qualified" definitions of the term "health" in *The Normal and the Pathological*: "Health, taken absolutely, is a normative concept defining an ideal type of organic structure and behavior; in this sense it is a pleonasm to speak of good health because health is organic well-being. Qualified health is a descriptive concept, defining an individual organism's particular disposition and reaction with regard to possible diseases" (137).

11. See the essays in Anne Fagot-Largeault, Claude Debru, Michel Morange, and Hee-Jin Han, eds., *Philosophie et médecine, en hommage à Georges Canguilhem* (Paris: Vrin, 2008), and in Pierre F. Daled, ed., *Alentour de Canguilhem: L'envers de la raison* (Paris: Vrin, 2009).

12. The historian of biology Michel Morange has argued that *The Normal and the Pathological* betrays Canguilhem's ignorance of then-contemporary biology—an ignorance that, for Morange, blunts his criticism. One could perhaps reverse this point and insist on the proximity between Canguilhem's criticism and the transformations in biology. See Morange, "Retour sur le normal et le pathologique," in *Philosophie et médecine*, 159.

13. One of Canguilhem's favorite quotes concerned the writing of the philosophy of history and the history of science: Koyré had declared, in the *Bulletin de la Société française de philosophie* 36, no. 3 (May–June 1936): "The history of science is certainly not a dead history. Nevertheless it is, *grosso modo*, the history of dead things." This passage, which Canguilhem first quoted in a March 1945 lecture "The History of Philosophy and the History of Science" (CAPHES [ENS], Archives de Georges Canguilhem, GC.11.3.10), formed the basis for some of his own reflections on writing the history of science, for example in "Cell Theory," where he asked "If the true—the goal of scientific research—is exempt from historical transformation, then is the history of science anything more than a museum of errors of human reason?" and sought to demonstrate the insufficiency of scientistic reductions of this sort and the legitimacy of his historical endeavor. See Canguilhem, *Knowledge of Life*, 26.

14. Jean Cavaillès, *Sur la logique et la théorie de la science* (Paris: Presses Universitaires de France, 1947).

15. See Canguilhem's essay on Cavaillès, republished as Georges Canguilhem, *Vie et mort de Jean Cavaillès* (Paris: Editions Allia, 1996); Gaston Bachelard, *Études*, ed. and with an introduction by Georges Canguilhem (Paris: Vrin, 1970); and Canguilhem's foreword to Dominique Lecourt, *L'epistemologie historique de Gaston Bachelard* (Paris: J. Vrin, 1969).

16. Foucault's oft-quoted homage to Canguilhem, "Life: Experience and Science," is published in English as the "Introduction" to *The Normal and the Pathological*, 7–24.

17. See Jean-François Braunstein, "Deux philosophies de la médecine: Canguilhem et Fleck," in Fagot-Largeault et al, *Philosophie et médecine*, 63–80.

18. Emile Durkheim, *The Rules of Sociological Method*, trans. W. D. Halls (New York: Free Press, 1982), chap.3. Maurice Halbwachs, *Théorie de l'homme moyen: Essai sur Quetelet et la statistique morale* (Paris, F. Alcan, 1913).

19. In France, it was customary to submit a "major" and a "minor" doctoral thesis.

20. See "Machine and Organism" in Canguilhem, *Knowledge of Life*, 75–97.

21. On Canguilhem and Bergson, see Georges Canguilhem and Gilles Deleuze, *Il significato della vita: Letture del III capitolo dell'Evoluzione creatrice di Bergson*, ed. Giuseppe Bianco (Milan: Mimesis Edizioni, 2006); Frédéric Worms, "La conception du vivant comme philosophie première" in Fagot-Largeault et al, *Philosophie et médecine*, 142–44; and the very detailed account in Giuseppe Bianco, "La réaction au bergsonisme: Transformations de la philosophie française de Politzer à Deleuze" (Ph.D. thesis, Université de Lille–III, 2010), chap. 4. To understand the trickiness of refusing mechanism at the time and the pitfalls of classical vitalism that Canguilhem was negotiating and trying to avoid, see Ernst Cassirer's contemporary essay "The Argument over Vitalism," in *The Problem of Knowledge: Philosophy, Science, and History Since Hegel*, (New Haven: Yale University Press, 1969), 188–216.

22. Canguilhem, "What is a Scientific Ideology?" in *Ideology and Rationality in the History of the Life Sciences*, 27–40.

23. Maurice Merleau-Ponty, *La nature: Notes, cours du Collège de France*, ed. Dominique Séglard (Paris: Seuil, 1995); *Nature: Course Notes from the Collège de France*, trans. Robert Vallier (Evanston, Ill.: Northwestern University Press, 2003).

24. Edmund Husserl, *Die Krisis der europäischen Wissenschaften und die transzendentale Phänomenologie: Eine Einleitung in die phänomentologische Philosophie*, ed. Walter Biemel (The Hague: Martinus Nijhoff, 1954); translated as *The Crisis of European Sciences and Transcendental Phenomenology: An Introduction to Phenomenological Philosophy*, trans. David Carr (Evanston: Northwestern University Press, 1970).

25. It is worth noting that such an idea differs profoundly from Erwin Schrödinger's *What Is Life?: The Physical Aspect of the Living Cell* (Cambridge: Cambridge University Press, 1945) just as it does from André Lwoff's *L'ordre*

biologique: Une synthèse magistrale des mècanismes de la vie (Verviers: Marabout Université, 1962) or François Jacob's, in *The Logic of Life*

26. Canguilhem, *Knowledge of Life*, 41.

27. Rudolf Virchow, *Die Cellularpathologie, in ihrer Begründung auf physiologischer und pathologischer Gewebelehre* (Berlin: Verlag von August Hirschwald, 1858).

28. This is the principal reason why Foucault repeatedly stated that his critiques of rationalism in his *History of Madness* and of the clinical gaze in *The Birth of the Clinic* were intellectual offspring of *The Normal and the Pathological* and *Knowledge of Life*. See Michel Foucault, *Histoire de la folie* (Paris: Plon, 1961); translated as *History of Madness*, ed. Jean Khalfa, trans. Jonathan Murphy and Jean Khalfa (London: Routledge, 2007); and Michel Foucault, *Naissance de la clinique* (Paris: Presses Universitaires de France, 1963); translated as *The Birth of the Clinic*, trans. A. M. Sheridan Smith (New York: Vintage, 1994).

29. Dominique Lecourt, *Georges Canguilhem* (Paris: Presses Universitaires de France, 2008), 5.

30. Lecourt broaches this theme clearly, but indirectly: Lecourt, *Georges Canguilhem*, 70, 83, 88–89.

31. Husserl's treatment of modern science as emergent in the Renaissance rediscovery of Greek universal science and as fundamentally rigidifying and indeed distorting human wonder at the world comes close to Canguilhem's concern with mechanism. Canguilhem's critique of psychology operates in a somewhat different register from Husserl's attack on psychologism, but the refusal of empiricism and the accusation of antiphilosophical intent are shared (or, rather, echoed) by Canguilhem. See also David Hyder, "Foucault, Cavaillès, and Husserl on the Historical Epistemology of the Sciences," *Perspectives on Science* 11, no. 1 (2003): 107–29. Adorno's negative dialectics, his rethinking of the particular in the 1950s and 1960s, and his ambivalent relationship with Kant could also be seen as similar to Canguilhem's reconceptualization of the individual, to the degree that both aimed to defend this individual or particular against totalitarianisms and totalizing systems and philosophy. As for Arendt, Canguilhem's negative vitalism is premised less on the affirmation of life and more on the consistent refusal—political, philosophical, and ethical—to efface life as creativity and absolute originality. It is in this sense that Canguilhem's argument can be read as closely related to Arendt's in *The Origins of Totalitarianism*, where against totalitarianism Arendt concludes by invoking "natality"—the birth of every new individual human being—as both a sign of what totalitarianism seeks to efface and a mark of what resists it.

32. The book plan can be found in "Corps et santé," CAPHES (ENS), Archives de Georges Canguilhem, GC.25.26, 21.

33. Chapter 1 in the present volume, "L'idée de nature" was originally published in *Médecine de l'homme, Revue du Centre catholique des médecins français*, no. 43 (March 1972): 6–12.

34. Chapter 4 in the present volume, "Une pédagogie de la guérison est-elle possible?" was originally written at the invitation of J.-B. Pontalis and published in *Nouvelle revue de psychanalyse* 17 (Spring 1978): 13–26. In 2006, a translation of this essay by Steven Miller appeared in *Umbr(a)*, a journal published by the Center for the Study of Psychoanalysis and Culture at the State University of New York, Buffalo. While we respect the work carried out in that version, we chose to retranslate the essay so as to maintain a single voice throughout the volume.

35. Chapter 3 in the present volume, "La santé: Concept vulgaire et question philosophique" originally appeared in *Cahiers du séminaire de philosophie*, Éditions du Centre de documentation en histoire de la philosophie 8 (1988):119–33. Our earlier translation of the essay was previously published in *Public Culture* 20, no. 3 (2008), 467–77; the current translation has been modified substantially from that version; significantly, we have chosen to revisit our rendition of *vulgaire* and render it as "popular."

36. Chapter 2 in the present volume, "Les maladies," appeared in the *Encyclopédie philosophique universelle*, vol. 1, *L'univers philosophique*, ed. André Jacob (Paris: Presses Universitaires de France, 1989), 1233–36. Our earlier translation of the essay was published in Richard Baxstrom and Todd Meyers, eds. *anthropologies* (Baltimore: Creative Capitalism Press, 2008); the current translation amends that version.

37. Georges Canguilhem, "Puissance et limites de la rationalité en médecine," in *Études d'histoire et de philosophie des sciences*, 392–411. For "Pour les dentistes," of which only a fragment survives, see CAPHES (ENS), Archives de Georges Canguilhem, GC.25.29.

38. Chapter 5 in the present volume, "Le problème des régulations dans l'organisme et dans la société" originally appeared in *Cahiers de l'Alliance israelite universelle*, no. 92 (September–October 1955): 64–73.

39. The idea of "the body proper" as a given, unchanged object in relation to contemporary biomedical thought, as discussed by Margaret Lock and Vinh-Kim Nguyen in *An Anthropology of Biomedicine* (London: Wiley-Blackwell, 2010), was particularly useful here.

40. Canguilhem foregrounds this through examples from endocrinology and immunology in Chapter 2, "Diseases."

41. See Jean-François Braunstein, "Psychologie et milieu: Ethique et histoire des sciences chez Georges Canguilhem" in Jean-François Braunstein, ed., *Canguilhem: Histoire des sciences et politique du vivant* (Paris: Presses Univversitaires de France, 2007).

42. See, for example, his references in this volume to the work of the medical historians Erwin H. Ackerknecht and Charles Daremberg.

43. The reference to a "Platonist" history of ideas concerns the tradition best known through Arthur O. Lovejoy's magnum opus *The Great Chain of*

Being: A Study of the History of an Idea (Cambridge: Harvard University Press, 1933).

44. As a method and rubric, "historical epistemology" has been reclaimed and practiced in recent years—and often with explicit reference to Canguilhem—by such scholars as Lorraine Daston, Ian Hacking, Arnold Davidson, and Hans-Jörg Rheinberger. See Arnold I. Davidson, *The Emergence of Sexuality: Historical Epistemology and the Formation of Concepts* (Cambridge, MA: Harvard University Press, 2001); Lorraine Daston and Peter Galison, *Objectivity* (New York: Zone Books, 2008); Ian Hacking, *Historical Ontology* (Cambridge, MA: Harvard University Press, 2002); Hans-Jörg Rheinberger, *An Epistemology of the Concrete: Twentieth-Century Histories of Life* (Durham, N.C.: Duke University Press, 2010).

45. Rheinberger, *On Historicizing Epistemology* (Stanford: Stanford University Press, 2010), 66.

46. Canguilhem, "L'objet de l'histoire des sciences," in *Études d'histoire et de philosophie des sciences*, 18.

47. Ivan Illich, *Medical Nemesis: The Expropriation of Health* (New York: Pantheon, 1982). See also René Dubos, *Man and His Environment: Biomedical Knowledge and Social Action* (Washington, D.C.: Pan-American Health Organization, March 1966), also cited in Illich's *Medical Nemesis* as a major source for understanding the role of the individual in defining health; Owsei Temkin, *The Double Face of Janus and Other Essays in the History of Medicine* (Baltimore: Johns Hopkins University Press, 2006); and Temkin, *Galenism: Rise and Decline of a Medical Philosophy* (Ithaca: Cornell University Press, 1973).

48. The illusion of physician competence and the political transmission of medical authority (and the danger of iatrogenic pathology) are polemics for which Canguilhem has little patience. See Erwin H. Ackerknecht's historical essay on "doctor-produced" illness, "Zur Geschichte der iatrogenen Krankheiten," *Gesnerus: The Swiss Journal for the History of Medicine and the Sciences* 27 (1970): 57–63.

49. Canguilhem's most notable critique of psychology is "Qu'est-ce que la psychologie?" in *Études d'histoire et de philosophie des sciences*. It should be mentioned that Canguilhem's library shows an unexpectedly broad interest in psychoanalysis.

50. Georges Canguilhem, "Nature naturante et nature dénaturée," in *Savoir, faire, espérer: Les limites de la raison*, vol.1 (Brussels: Publications des Facultés Universitaires Saint-Louis, 1976), 74 and 71.

51. Ibid., 87.

52. Canguilhem, *The Normal and the Pathological*, from the chapter "Disease, Cure, Health," p. 187.

53. See, for instance, the discussion by Michel Foucault of a philosophical anthropology that is aimed not at discovering what is hidden, but is meant to

"render visible what precisely is visible." Michel Foucault, *Dits et écrits, 1954–1988*, vol. 1 (Paris: Gallimard, 2001), 540–41.

54. See Henning Schmidgen, "Leben und Erkenntnis: Über eine Entwicklung im Werk von Georges Canguilhem," *Paragrana* 17, no. 2 (December 2008): 33–43.

55. Friedrich Nietzsche, *Beyond Good and Evil*, trans. Walter Kaufmann (New York: Vintage Books, 1966), 153–55.

56. Canguilhem, *The Normal and the Pathological*, 197.

57. Walter B. Cannon, "The Interrelations of Emotions as Suggested by Recent Physiological Researches," *American Journal of Psychology* 25 (1914): 256.

58. Hans Selye, "Syndrome Produced by Diverse Nocuous Agents," *Nature* 138 (July 4, 1936): 32. See Selye, *The Stress of Life* (1956; New York: McGraw Hill, 1976). See also Pierre A. Buser and Michel Imbert, *Neurophysiologie fonctionnelle* (Paris: Hermann, 1975), 417.

59. See, for example, René Thom, *Structural Stability and Morphogenesis: An Outline of a General Theory of Models*, trans. D. H. Fowler (Reading, MA.: W. A. Benjamin, 1975).

60. Canguilhem, *The Normal and the Pathological*, 182–83; Kurt Goldstein, *The Organism*, with a foreword by Oliver Sacks (New York: Zone Books, 1995), 276–77.

61. Canguilhem, "Thérapeutique, expérimentation, responsabilité," in *Études d'histoire et de philosophie des sciences*, 383–91.

62. Ibid., 391.

63. See, Georg Groddeck, *The Book of the It*, trans. V. M. E. Collins (New York: Vintage, 1949).

64. René Allendy, *Essai sur la guérison* (Paris: Denoël & Steele, 1934). See Anaïs Nin, *The Diary of Anaïs Nin, vol. 1: 1931–1934* (Orlando: Harcourt Books, 1994). See also the discussion of Allendy's influence of Canguilhem in Lecourt, *Canguilhem*, 34–36.

65. Goldstein, *The Organism*, 341.

66. Canguilhem, "Concepts biologiques fondamentaux au XIXe siècle: Milieu intérieur et régulations, 1957–58," in CAPHES (ENS) GC.12.2.4.

67. Canguilhem, s.v. "Régulation (Épistémologie)," *Encyclopaedia Universalis*, vol. 14 (Paris: Encyclopaedia Universalis—France, 1972), 1–3.

68. On the brain, it is important to also note Canguilhem's famous article "Le cerveau et la pensée" where Canguilhem struggled against neurology's theorization of the brain as more or less secreting thought. See É. Balibar, M. Cardot, F. Duroux, M. Fichant, D. Lecourt, and J. Roubaud, eds., *Georges Canguilhem, philosophe, historien des sciences: Actes du colloque (6–8 décembre 1990* (Paris: Albin Michel/Collège International de Philosophie, 1993), 11–33.

69. Canguilhem, "La régulation comme réalité et comme fiction" in CAPHES (ENS) Archives de Georges Canguilhem, GC.25.17, 93.

70. Walter B. Cannon, *The Wisdom of the Body*, 2nd ed. (1932; New York: W. W. Norton, 1963); Canguilhem, "La régulation comme réalité et comme fiction" in CAPHES (ENS) GC.25.17, 95–96 and 98.

71. "La formation du concept biologique de regulation aux *XVIIe et XVIIIe* siècles" was first published in *L'idée de regulation dans les sciences*, Séminaires interdisciplinaires du Collège de France 2 (Paris: Maloine, 1977).

72. Canguilhem, "Régulation 1974," in CAPHES (ENS), Archives de Georges Canguilhem, GC.17.3.

1. THE IDEA OF NATURE IN MEDICAL THEORY AND PRACTICE

1. [According to Hooper's *Lexicon-Medicum* of 1826, *stenia* is defined as "From σθένος: a term employed by the followers of Dr. Brown, to denote that state of the body which disposes to inflammatory diseases in opposition to those of debility, which arise from asthenia"; and *asthenia* is defined as "From *a*, priv. and σθένος: Extreme debility. The asthenic diseases form one great branch of Brunonian arrangement." Robert Hooper, *Lexicon-Medicum* (New York: J. & J. Harper, 1826), 901, 115.—Trans.]

2. John Brown, *The Elements of Medicine; or, A Translation of* the "Elementa medicinae Brunonis" (Philadelphia: Thomas Dobson, 1795), §95, 127, trans. modified. [The full quote is as follows: "Since every universal disease, every predisposition, depends upon increased or diminished excitement, and is removed by the conversion of that into the degree which constitutes the mean betwixt both; in order to both prevent and cure diseases, we must always use the indication proposed, and stimulate or debilitate; never wait, or trust to the supposed powers of nature, which have no real existence." Canguilhem elaborates on Brown's work in "John Brown's System: An Example of Medical Ideology," in *Ideology and Rationality in the History of the Life Sciences*, trans. Arthur Goldhammer. (Cambridge: MIT Press, 1988), 41–50 and 55–56.—Trans.]

3. Ibid., §72, 116. [Canguilhem's rendition is inexact; Brown writes: "life is not a natural, but a forced state; that the tendency of animals every moment is to dissolution; that they are kept from it, not by any powers in themselves, but by foreign powers, and even by these with difficulty, and only for a time; and then, from the necessity of their fate they yield to death."—Trans.]

4. [The French term *expectant* has no simple English cognate. Svetolik P. Djordjević's *Dictionary of Medicine*, 2nd ed. (Rockville: Schreiber Publishing, 2000), 524, renders *médecine expectante* as "expectant medicine (expectant symptomatic treatment)"; the quotidian use of the French adjective *expectant* would be "attentive," "wait-and-see."—Trans.]

5. Théophile de Bordeu, *Recherches sur l'histoire de la médecine* (1768), in *Oeuvres completes de Bordeu*, ed. M. le Chevalier Richerand, vol. 2 (Paris: Caille et Ravier, 1818), 597.

6. Hippocrates, *The Art*, 4:1–5, 5:120, in *Hippocrates: Volume II*, trans. W. H. S. Jones (Cambridge, MA: Harvard University Press, 1992), 194–95, and 196–97.

7. Hippocrates, *Epidemics*, 6:5.1, in *Hippocrates: Volume VII*, ed. and trans. Wesley D. Smith, (Cambridge, MA: Harvard University Press, 1994), 254–55, trans. modified.

8. Hippocrates, *Epidemics*, 6:5.1, 254–55.

9. Hippocrates, *The Art*, 8:10–13, 202–3.

10. [Edouard Rist, *Histoire critique de la médecine dans l'Antiquité* (Paris: Les Amis d'Edouard Rist, 1966).—Trans.]

11. François Dagognet, *La raison et les remèdes: Essai sur l'imaginaire et le réel dans la thérapeutique contemporaine* (Paris: Presses Universitaires de France, 1964).

12. [Canguilhem is referring to a reaction of the immune system wherein an infection or injury causes an increase in proteolytic enzymes to neutralize the biochemical causing inflammation, allowing repair to take place.—Trans.]

13. [Canguilhem takes the term "pathogenic situations" from Hans Selye's *The Stress of Life* (1956; New York: McGraw Hill, 1976), 316.—Trans.]

14. [Hans Selye uses the concept of a "pathogenic situation" to describe "non-specific stress," a condition in which multiple factors precipitate disease. See Selye, *The Stress of Life*. But the term is in broader use: Sigmund Freud uses it in a somewhat different way in "The Origin and Development of Psychoanalysis," *American Journal of Psychology* 21(1910): 181–218: "So we are forced to the conclusion that the patient fell ill because the emotion developed in the pathogenic situation was prevented from escaping normally, and that the essence of the sickness lies in the fact that these "imprisoned" (*dingeklemmt*) emotions undergo a series of abnormal changes."—Trans.]

15. [Jean de La Fontaine, "Epistle to Huet" (1687): "Mon imitation n'est pas un esclavage, je ne prends que l'idée et les tours, et les loix." *Oeuvres Complètes de La Fountaine, Publiées d'après les textes originaux par Ch. Marty-Laveaux*, vol. 5, *Poesies Diverses* (Paris: Paul Daffis, 1877), 177.—Trans.]

16. Henry Dale, *Adventures in Physiology, with Excursions into Autopharmacology*. (London: Pergamon, 1953); see the sections "Histamine Shock" (with P. P. Laidlow), 290–319; "The Biological Significance of Anaphylaxis," 320–37; and "Conditions Which Are Conducive to the Production of Shock by Histamine," 338–48.

17. [See especially the final chapter, "The General Features of Bodily Stabilization," in Walter B. Cannon, *The Wisdom of the Body* (1932; New York: W. W. Norton, 1963), 286–304.—Trans.]

18. Ibid., 240.

19. Ibid., 242.

20. Max Neuburger, *Die Lehre von der Heilkraft der Natur im Wandel der Zeiten* (Stuttgart: Enke, 1926), trans. Linn J. Boyd as *The Doctrine of the Healing*

Power of Nature throughout the Course of Time (New York: New York Homeopathic College, 1932).

21. Evelyne Aziza-Shuster, *Le Médecin de soi-même* (Paris: Presses Universitaires de France, 1972).

22. Michel de Montaigne, "Of Experience," in *The Complete Essays of Montaigne*, trans. Donald Murdoch Frame (Stanford: Stanford University Press, 1958), 833.

23. [René Descartes, "Conversation with Burman, 16th of April, 1648" in *The Philosophical Writings of Descartes*, vol. 3, *The Correspondence*, ed. John Cottingham, Robert Stoothoff, Dugald Murdoch, and Anthony Kenny (Cambridge: Cambridge University Press, 1991), 354. In the "Conversation," Descartes also cites Tiberius on the "thirty years."—Trans.]

24. Jean Devaux, *Le Médecin de soi-même ou l'Art de conserver la santé par l'instinct* (Leyde: Chez de Graef, 1682).

25. John Archer, *Every Man His Own Doctor* (London: Peter Lillicrap, 1671, 1673).

26. [Georg Ernst Stahl, *De medicina sine medico* (Halle, 1707); Stahl, *De autocratia naturae siu spontanea morborum excusione* (Halle, 1696). Canguilhem's reference can use some unpacking. According to Hubert Steicke, in Théophile de Bordeu's *Recherches anatomiques sur la position des glandes et leur action* (1751), "the living animal as a whole was constituted by a general tonic movement in all parts of the body, which depended on a continuous irritation of the nerves through movements of the brain." Also, Stahl's doctrine on respiration describes the "vital tonic movement" as taking place "independently" and "quite apart from our will and consciousness." Hubert Steinke, *Irritating Experiments: Haller's Concept and the European Controversy on Irritability and Sensibility, 1750–90* (Amsterdam: Rodopi, 2005), 203–4.—Trans.]

27. [Friedrich Hoffmann, *De medico sui ipsius* (Halle, 1704). —Trans.]

28. [In the Galenic tradition, the six non-natural things were air, food and drink, sleep and waking, movement and rest, excretion and retention, and the emotions, or passions of the soul.—Trans.]

29. [Linnaeus was concerned with an etiology of disease that followed a line of thought found in Hippocratic medicine (specifically, the four seasons and four humors), what he called the "hot and cold," and "proximal" (change in tension in fluid and liquid parts) and "remote" (changes in diet, contagion, bodily constitution) causes of disease.—Trans.]

30. [Canguilhem is referring here to the long history of "inspection" of the body, which includes forms of "auscultation" (listening to the internal sounds of the body, of respiration, blood flow, abdominal movement), and "percussion" (tapping the body to listen for the resonance of masses in the body). Other forms not discussed here include "palpation" (touching, manipulating) and forms of visual inspection. —Trans.]

31. [Canguilhem is implicitly referring to Foucault's *The Birth of the Clinic*, trans. A. M. Sheridan (New York: Vintage, 1973), and in particular chapters 6, "Signs and Cases," and 7, "Seeing and Knowing."—Trans.]

32. Jean-Martin Charcot, *De l'expectation en médecine: Thèse de concours pour l'agregation présentée et soutenue à la Faculté de Médecine le 17 Avril 1857* (Paris: Librairie de Germer Baillière, 1857).

33. Emile Littré, "De l'hygiène," in *Médecine et médecins* (Paris: Didier, 1872).

34. [For a discussion of Ehrlich's work with Robert Koch and Emil von Behring, see chapter 4 of Martha Marquardt, *Paul Ehrlich*, with an introduction by Henry Dale (New York: Schuman, 1951); Reinhard Mocek, *Wilhelm Roux, Hans Driesch: zur Geschichte der Entwicklungsphysiologie der Tiere ("Entwicklungsmechanik")* (Jena: G. Fischer, 1974); Emil Adolf von Behring, *Beiträge zur experimentellen Therapie* (1906), cited in Charles Singer, *A History of Biology to About the Year 1900: A General Introduction to the Study of Living Things* (1931; London: Abelard-Schuman, 1956), 456.—Trans.]

35. [See Ehrlich's experiments in staining techniques and the treatment of diseases caused by protozoa, spirilla, and bacteria in Paul Ehrlich, *Collected Papers, Including a Complete Bibliography*, ed. Henry Dale (New York: Pergamon Press, 1956). —Trans.]

36. Georg Groddeck, *Natura sanat, medicus curat: Der gesunde und kranke Mensch* (Leipzig: Hirzel, 1913), also known as *Nasamecu*, translated into French as *"Nasamecu": La nature guérit* (Paris: Aubier Montaigne, 1980).

2. DISEASES

1. [Denis Diderot, *Essais sur la peinture: Salons de 1759, 1761, 1763*, ed. Jacques Chouillet (Paris: Hermann, 1984); ed. and trans. John Goodman as *Diderot on Art, The Salon of 1765 and Notes on Painting* (New Haven: Yale University Press, 1995), 191, trans. modified.—Trans.]

2. [Following Diderot, Canguilhem is referring here to Leibniz's formulation of the principle of reason: "nihil est sine ratione"—nothing is without reason. It is significant that Canguilhem does not cite the principle of reason, but instead, in a consideration of biology and an imposition of it onto the realm of life and disease, cites an *aesthetic* consideration premised on it. —Trans.]

3. [Hippocrates, *Aphorisms of Hippocrates, and the Sentences of Celsus*, trans. C. J. Sprengell (London: R. Bonwick, 1708). —Trans.]

4. [Plato, *Phaedrus*, 270c.10; trans. Alexander Nehamas and Paul Woodruff (New York: Hackett, 1995), 22, trans. modified. See also Wesley D. Smith, *The Hippocratic Tradition* (Ithaca, NY: Cornell University Press, 1979), 44–48. —Trans.]

5. Charles Nicolle, *Naissance, vie et mort des maladies infectieuses* (Paris: F. Alcan, 1930).

6. [Canguilhem seems to be referring to two separate threads of research on the function of enzymes. The first refers to research on the toxic effects of phytohaemagglutinin (PHA) on the immune system after the consumption of raw beans, including fava beans (*Vicia faba*), which contain a-amylase inhibitors. The second example refers to research on the role of glucose-6-phosphate dehydrogenase (G6PD, an enzyme) in genotype deficiencies in the protection against *Plasmodium falciparum* malaria in Africa.—Trans.]

7. Jacques Tenon, *Memoirs on Paris Hospitals*, ed. and trans. Dora B. Weiner (Canton, MA: Science History Publications, 1996).

8. Étienne Tourtelle, *The Principles of Health: Elements of Hygiene, or, A Treatise on the Influence of Physical and Moral causes on Man, and on the Means of Preserving Health* (Baltimore: John D. Toy, 1819).

9. Louis-René Villermé, *Tableau de l'état physique et moral des ouvriers employés dans les manufactures de cotton, de laine et de soie*, 2 vols. (Paris: Jules Renouard, 1840).

10. Hans Selye, *The Stress of Life* (1956; New York: McGraw-Hill, 1976).

11. [On the *francs-tireurs* group, see, for example, *Autrement* 9 (1977), special issue, *Francs-tireurs de la médecine: Dans leur pratique, que remettent en cause et que re-inventent ces "equipes" de santé?* —Trans.]

12. Thomas Sydenham, *Observationes medicae* (London, 1676), ed. G. G. Meynell (Folkestone, UK: Winterdown, 1991).

13. Sigmund Freud, letter to Lou Andreas-Salome, May 13, 1924, in Sigmund Freud and Lou Andreas-Salome, *Letters*, ed. Ernest Pfeiffer (New York: W.W. Norton, 1972), 135.

14. Freud, letter to Lou Andreas-Salome (May 10, 1925), in Freud and Andreas-Salome, *Letters*, 154.

15. E. H. Ackernecht, *History and Geography of the Most Important Diseases* (New York: Hafner, 1965), 70; François Dagognet, "Pour une historie de la médicine," in *Philosophie de l'image* (Paris: Vrin, 1984); François Dagognet, "Autopsie et tableau," in *Le nombre et le lieu* (Paris: Vrin, 1984); Mirko Drazen Grmek, *Les maladies à l'aube de la civilisation occidentale* (Paris: Payot, 1983).

16. [Canguilhem closed the final draft of his essay (see École Normale Supérieure, CAPHES, Archives de Georges Canguilhem, GC.26.2.13) with three extracts that were not included in the published version in the 1989 *Encyclopédie philosophique universelle*, from which the subsequent French edition of *Écrits sur la médecine* derives. We restore them here. The first is from Michel de Montaigne, *The Complete Essays of Montaigne*, trans. Donald M. Frame (Stanford: Stanford University Press, 1958), book 3, chapter 13, "Of Experience," 834.—Trans.]

17. [Claude Bernard, *Leçons sur le diabète et la glycogenèse animale* (Paris, Baillière 1877), 56–57.—Trans.]

18. [Michel Foucault, *The Birth of the Clinic*, trans. A. M. Sheridan (New York: Vintage, 1973), 155.—Trans.]

3. HEALTH: POPULAR CONCEPT AND PHILOSOPHICAL QUESTION

NOTE: [As noted in the Introduction, the text of "La santé: Concept vulgaire et question philosophique," a lecture Canguilhem gave in Strasbourg in 1988, is a shortened version of another lecture he had delivered more than a decade earlier at the Université Paris–I. In the notes to the present chapter, we have included a number of the citations that Canguilhem makes in that lecture, "Corps et santé," because it includes far more elaborate citations. Some of these references are indeed remarkable, because they address philosophers and movements with which Canguilhem has not been known to engage explicitly, notably Wittgenstein, Heidegger, and structuralism. It should nevertheless be remembered that, despite presenting the lecture "Corps et santé" in a public forum, Canguilhem did not approve it for publication and that these references are offered here as indications of the conversations he held.—Trans.]

1. [Epictetus, *Dissertationes ab Arriano digestae*, 2.17.8.—Trans.]

2. [The essay was originally given as a lecture in Strasbourg in May 1988.—Trans.]

3. [Canguilhem is referring to René Leriche. See Canguilhem's discussions of Leriche in *The Normal and the Pathological*, trans. Carolyn Fawcett (New York: Zone, 1989), 93–101, and in *Knowledge of Life*, ed. Paola Marrati and Todd Meyers, trans. Stefanos Geroulanos and Daniela Ginsburg (New York: Fordham University Press, 2008), 129–31.—Trans.]

4. [Paul Valéry, *Mauvaises pensées et autres* (Paris: Gallimard, 1942), 191.—Trans.]

5. Charles Daremberg, *La médecine, histoire et doctrines*, 2nd ed. (Paris: Didier, 1865).

6. [Henri Michaux, *Les grandes épreuves de l'esprit et les innombrables petits* (Paris: Gallimard, 1966), 14.—Trans.]

7. [Denis Diderot, *Lettre sur les aveugles; Lettre sur les sourds et muets* (Paris: Flammarion, 2000), 110, translated by Margaret Jourdain in Diderot, *Early Philosophical Writings* (New York: Lenox Hill, 1916), 185–86, translation amended.—Trans.]

8. Gottfried Wilhelm Leibniz, *Theodicy*, trans. E. M. Huggard (Indianapolis: Bobbs-Merrill, 1966), §251, 281.

9. Ibid., §259, 285–86.

10. Andreas Christoph Wasianski, *Immanuel Kant, sein Leben in Darstellungen von Zeitgenossen* (Berlin: Deutsche Bibliothek, 1912).

11. Immanuel Kant, *The Conflict of the Faculties*, trans. Mary J. Gregor (New York: Abaris, 1979), 181.

12. Descartes to Hector Pierre Chanut, March 31, 1649, in *The Philosophical Writings of Descartes*, trans. John Cottingham, Robert Stoothoff, and Dugald Murdoch, vol. 3 (Cambridge: Cambridge University Press, 1991), 370.

13. [In his lecture "Corps et santé" (April 29, 1977), Canguilhem notes that Heidegger "defines truth as nonveiling and exactitude."—Trans.]

14. [In "Corps et santé," Canguilhem defends his etymological effort through philosophical references that do not appear in the published version of the "Health" essay, citing structuralism and Wittgenstein when he writes: "We always learn when we seek to resuscitate, beneath the dust of significations and beneath the banality of uses ('Wittgenstein used to say that words don't have meanings but only uses'), the relation, lived in a given situation, between a signifier and a signified." Canguilhem further quotes Georges Bataille's point in "Formless" that "a dictionary begins when it no longer gives the meanings of words but their tasks." Georges Bataille, *Visions of Excess: Selected Writings, 1927–1939*, trans. Allan Stoekl (Minneapolis: University of Minnesota Press, 1985), 31. See ENS/CAPHES, Archives de Georges Canguilhem, GC.25.26, "Corps et santé," 2.—Trans.]

15. Édouard Brissaud, *Histoire des expressions populaires relatives à l'anatomie, à la physiologie et à la médecine*, 2nd ed. (Paris: Chamerot, 1892).

16. [Charles Andler, *Nietzsche, sa vie et sa pensée*, 3 vols. (Paris: Gallimard, 1958). —Trans.]

17. [Ernst Bertram, *Nietzsche: Versuch einer mythologie* (Berlin: G. Bondi, 1918).—Trans.]

18. [Karl Jaspers, *Nietzsche: An Introduction to the Understanding of His Philosophical Activity* (1965; Baltimore: Johns Hopkins University Press, 1997). —Trans.]

19. [Karl Löwith, *From Hegel to Nietzsche: The Revolution in Nineteenth-Century Thought*, trans. David E. Green (1964; New York: Columbia University Press, 1991); Löwith, *Nietzsche* (Stuttgart: Metzler, 1987).—Trans.]

20. [Friedrich Nietzsche, *The Will to Power*, trans. Walter Kaufmann and R. J. Hollingdale (New York: Vintage, 1967), §47, 29–30.—Trans.]

21. [Friedrich Nietzsche, *The Gay Science*, trans. Walter Kaufmann (New York: Random House, 1974), §382, 346.—Trans.]

22. Friedrich Nietzsche, *Twilight of the Idols and The Anti-Christ*, trans. R. J. Hollingdale (New York: Penguin, 1990), 180. [We translate directly from Canguilhem. Hollingdale's translation is "everything well-constituted, proud, high-spirited, beauty above all." —Trans.]

23. [Friedrich Nietzsche, *Thus Spoke Zarathustra*, trans. Adrian del Caro (Cambridge: Cambridge University Press, 2006), 22.—Trans.]

24. [Ibid., 23. —Trans.]

25. Ernest Henry Starling, "The Wisdom of the Body: The Harveian Oration Delivered before the Royal College of Physicians of London on St. Luke's Day, 1923," *British Medical Journal* 3277 (October 20, 1923): 685–90; Walter B. Cannon, *The Wisdom of the Body* (1932; New York: W. W. Norton, 1963).

26. Ernest Henry Starling and C. Lovatt Evans, *Principles of Human Physiology* (London: Lea and Febiger, 1936).

27. Charles Kayser, *Physiologie*, vol. 3, *Les grandes fonctions* (Paris: Flammarion, 1992).

28. Claude Bernard, *Leçons sur le diabète et la glycogenèse animale* (Paris: Baillière, 1877), 72, 354, 421.

29. Starling and Evans, *Principles of Human Physiology*, p. 7. [The quotation is from the edition revised by Evans. In his original edition, referred to by Canguilhem, Starling also defied mechanism (and vitalism) by writing that "for [the physiologist] to describe himself as a vitalist or mechanist is as germane to the subject as if he were to call himself a Trinitarian or a Plymouth Brother." Ernest Henry Starling, *Principles of Human Physiology* (1912; London: Churchill, 1968), 8.—Trans.]

30. [René Descartes, *Meditations on First Philosophy*, in *The Philosophical Writings of Descartes*, vol. 2, ed. John Cottingham and Dugald Murdoch (London: Cambridge University Press, 1985), 58.—Trans.]

31. [Ibid., 59.—Trans.]

32. Raymond Ruyer, "Le paradoxe du régulateur parfait," in *Paradoxes de la conscience et limites de l'automatisme* (Paris: Albin Michel, 1966), 232.

33. Auguste Villiers de l'Isle-Adam, *Tomorrow's Eve*, trans. Robert Martin Adams (Urbana: University of Illinois Press, 2001), 203, trans. modified on the basis of Canguilhem's quotation. [In his lecture "Corps et santé," Canguilhem further cites Raymond Ruyer on the problem of the man as machine in *Paradoxes de la conscience*, 22.—Trans.]

34. Samuel Auguste André David Tissot, *Avis au peuple sur sa santé* (Paris: P. Fr. Didot le Jeune, 1782); Tissot, *De la santé des gens de lettres* (Geneva: Slatkine, 1981).

35. Étienne Tourtelle, *Éléments d'hygiène* (Strasbourg: Levrault, 1797); Tourtelle, *The Principles of Health: Elements of Hygiene; or, A Treatise on the Influence of Physical and Moral Causes on Man, and on the Means of Preserving Health*, trans. G. Williamson, from 2nd French ed., corr. and enl. (Baltimore: Toy, 1819). [In "Corps et santé," Canguilhem contextualizes his citation by suggesting that Tourtelle borrows from Thomas Sydenham's 1666 *Médecine pratique* (Paris: Théophile Barrois, 1784).—Trans.]

36. Constitution of the World Health Organization, Principle 1, WHO, Basic Documents, 45th ed., supplement (October 2006).

37. Antonin Artaud, "Lettre à la voyante," *La révolution surréaliste* (December 1, 1926), 17n1.

38. [Canguilhem is referring here to la Sécurité Sociale, or La Sécu, as it is popularly known, a four-branch form of government social protection providing social welfare for illness, old age, family, and recovery in France.—Trans.]

39. Nicolas Malebranche, "Dialogue III," in *Dialogues on Metaphysics and on Religion*, ed. Nicholas Jolley, trans. David Scott (Cambridge: Cambridge University Press, 1997), 34–35.

40. [Michel Henry, *Philosophie et phénoménologie du corps: Essai sur l'ontologie biranienne* (Paris: Presses Universitaires de France, 1965), 190 and 197; trans.

Girard Etzkorn as *Philosophy and Phenomenology of the Body* (The Hague: Martinus Nijhoff, 1975), 137 and 139.—Trans.]

41. Maurice Merleau-Ponty, *L'union de l'âme et du corps chez Malebranche, Biran et Bergson* (Paris: Vrin, 1968), translated by Paul B. Milan as *The Incarnate Subject: Malebranche, Biran, and Bergson on the Union of Body and Soul* (Amherst: Humanity Books, 2001). Merleau-Ponty, *Nature: Course Notes from the Collège de France*, comp. and ed. Dominique Séglard, trans. Robert Vallier (Evanston: Northwestern University Press, 2003).

42. Maurice Merleau-Ponty, *The Visible and the Invisible*, ed. Claude Lefort, trans. Alphonso Lingis (Evanston: Northwestern University Press, 1968), 234.

43. [In "Corps et santé," Canguilhem, having already noted and distanced himself from Michel Henry's severity toward Descartes, becomes far harsher toward Merleau-Ponty, whom he had praised as early as *The Normal and the Pathological*, writing that when compared with Descartes' philosophy, Merleau-Ponty's, despite its ambition, does not rise to the level of either an ontology or a phenomenology.—Trans.]

44. Maurice Merleau-Ponty, *Résumés de cours, Collège de France, 1952–1960* (Paris: Gallimard, 1968).

45. Ruyer, *Paradoxes de la conscience*, 285. I could not abstain here from evoking the late Roger Chambon. In his thesis, *Le monde comme perception et réalité* (Paris: Vrin, 1974), he brilliantly presented and discussed the works of Henry and Merleau-Ponty and more attentively still those of Ruyer.

46. [Canguilhem is referring to Ivan Illich, *Medical Nemesis: The Expropriation of Health* (New York: Pantheon Books, 1976).—Trans.]

47. [René Descartes, Letter to Princess Elizabeth, 28 June 1643, in *The Philosophical Writings of Descartes*, vol. 3, *Correspondence*, trans. John Cottingham, Robert Stoothoff, Dugald Murdoch and Anthony Kenny (Cambridge: Cambridge Univ. Press, 1991), 227.—Trans.]

48. Merleau-Ponty, *The Visible and the Invisible*, 27, trans. modified.

4. IS A PEDAGOGY OF HEALING POSSIBLE?

1. See François Dagognet's *La raison et les remèdes* (Paris: Presses Universitaires de France, 1964), in particular, chapter 1; Pierre Kissel and Daniel Barrucand, *Placebos et effet placebo en médecine* (Paris: Masson, 1964); Daniel Schwartz, Robert Flamant, and Joseph Lellouch, *L'essai thérapeutique chez l'homme* (Paris: Flammarion 1970).

2. "I have followed methods of medical treatment of every kind, at one time so, another time so, and have found that all roads lead to Rome, those of science and those of charlatanry." Georg Groddeck, *The Book of the It*, trans. V. M. E. Collins (New York: Vintage, 1949), 241. In his preface to the English-language translation of *The Book of the It*, Lawrence Durrell writes, "Groddeck was more of a healer and a sage than a doctor." [The translation of Durrell's sentence

back into English has been modified to echo Canguilhem's French edition, *Le livre du ça* (Paris: Gallimard, 1973). Durrell calls Groddeck, among other things, a "philosopher and healer," "poet-philosopher-doctor," "watchman," "natural philosopher," and "wily metaphysician."—Trans.]

3. René Allendy, *Essai sur la guérison* (Paris: Denoël et Steele, 1934), and, earlier still, *L'orientation actualle des idées médicales* (Paris: Au Sans Pareil, 1927). One could also cite, on the basis of his collaboration with Allendy, René Laforgue, *Clinique psychanalytique* (Paris: Denoël et Steele, 1936) [trans. Joan Hall as *Clinical Aspects of Psychoanalysis* (London: Hogarth, 1938)], particularly lecture 7, "Curing and the Completion of Treatment," which does not exclusively concern the psychoanalytic cure.

4. "One must not forget that recovery is brought about not by the physician, but by the sick man himself. The patient heals himself, by *his own power*, exactly as *he walks* by means of *his own power*, or *eats*, or *thinks*, *breathes* or *sleeps*." Groddeck, *The Book of the It*, 243.

5. See Yvon Belavel, *Les conduites d'échec* (Paris: Gallimard, 1953). [Trans.: Canguilhem's use of *accident* and *conduite* (behavior, conduct) here makes use of an untranslatable driving metaphor].

6. Dominique Raymond, *Traité des maladies qu'il est dangereux de guérir* (Avignon: E. B. Mirande, 1757), republished in an expanded edition with notes by M. Giraudy (Paris: Brunot Labbe, 1808).

7. Jean-Martin Charcot, *De l'expectation en médecine: Thèse de concours pour l'agregation présentée et soutenue à la Faculté de Médecine le 17 Avril 1857* (Paris: Librairie de Germer Bailliere, 1857).

8. See Charles Lichtenthaler, "De l'origine sociale de certains concepts scientifiques et philosophiques grecs," in *La Médecine hippocratique* (Neuchâtel: La Baconnière, 1957); Bernard Balan, "Premières recherches sur l'origine et la formation du concept d'économie animale," *Revue d'histoire des sciences* 28, no. 4 (1975): 289–326.

9. Leibniz, theoretician of the conservation of force, introduces as an argument into his system the Hippocratic theorem of the conservation of organic "forces" a point of agreement between rival doctors at the University of Halle—the animist Georg Ernst Stahl and the mechanist Friedrich Hoffmann: "I am not astonished men are sometimes sick, but I am astonished they are sick so little and not always. This also ought to make us esteem the divine contrivance of the mechanisms of animals, whose Author has made machines so fragile and so subject to corruption and yet so capable of maintaining themselves: for it is Nature which cures us rather than medicine." See Gottfried Wilhelm Leibniz, *Theodicy*, trans. E. M. Hugard (Indianapolis: Bobbs-Merrill, 1966), part 1, §14, 37.

10. [Oxyuriasis is an intestinal infection caused by small worms (pinworms). It is most common in children.—Trans.]

11. ["Constitution" here refers to Hippocrates' term *katastasis* (in the books of the *Epidemics*), taken up by Galen, which refers to external influence on the individual and is often used to describe a climate or environment; Thomas Sydenham rethought the Hippocratic understanding of a season or climate into a theory of epidemic constitutions and the milieus appropriate to them. —Trans.]

12. [Thomas Sydenham, *The Works of Thomas Sydenham, on Acute and Chronic Diseases, with Their Histories and Modes of Cure* (Philadelphia: Benjamin and Thomas Kite, 1809), xxiv.—Trans.]

13. Auguste Comte, *Course in Positive Philosophy*, lesson 40, in *The Positive Philosophy of Auguste Comte*, trans. Harriet Martineau, (1836; New York: D. Appleton and Co., 1853), 356–57]; Emile Gley. "La Société de biologie de 1849 à 1900 et l'évolution des sciences biologiques," in *Essais d'histoire et de philosophie de la biologie* (Paris: Masson, 1900), 187. Also see s.v. "Mésologie" in Emile Littré and Charles Robin, *Dictionnaire des sciences médicales* (Paris: Hachette, 1844–1852). [Canguilhem defines his conception of "anticipation" in the context of the history of science in *Knowledge of Life* (New York: Fordham University Press, 2008), 38–39.—Trans.]

14. See Jean Laplanche, "Why the Death Drive?" in *Life and Death in Psychoanalysis*, trans. Jeffrey Mehlman (Baltimore: Johns Hopkins University Press, 1976), 103–24. The author shows to what extent and in what way Freud relied, not without confusion, upon Hermann von Helmholtz's work on energetics. [Trans.: Canguilhem is referring to Sigmund Freud, *Beyond the Pleasure Principle*, ed. and trans. James Strachey (New York: W. W. Norton, 1989), chapters 5–6.]

15. [For an extended discussion of René Leriche's definition of health as "life lived in the silence of the organs," see Chapter 3 of the present volume. —Trans.]

16. Regarding the different concepts and evaluations of the cure, see Jacques Sarano, *La guérison* (Paris: Presses Universitaires de France, 1955).

17. The expression is from La Fontaine's *Fables*, book 9, fable 5: "The Sculptor and the Statue of Jupiter," trans. modified. [In Jean de La Fontaine, *The Complete Fables of Jean de La Fontaine*, trans. Norman R. Shapiro (Urbana: University of Illinois Press, 2007), it is translated thus: "Finding it fair and passing fine,/A sculptor bought a marble block./'What shall I sculpt, O chisel mine?/A god, a taboret, a clock?'" (240). —Trans.]

18. On the history of tuberculosis, see Marius Piéry and Julien Roshem, *Histoire de la tuberculose* (Paris: Doin, 1931) and Charles Coury, *La tuberculose au cours des âges* (Suresnes: Lepetit, 1972).

19. Jean-Bertrand Pontalis recognizes the ambiguity of the term "psychology," which designates both a discipline and its object, as if the representation of a self were already constitutive of the representing subject. See Jean-Bertrand Pontalis, *Entre le rêve et la douleur* (Paris: Gallimard, 1977), 135.

20. See, for example, Marie Bashkirtseff's diary (Thursday, December 28, 1882): "Potain never wanted to say that the lungs were affected; he would use

the usual formulas for such a case, the bronchial tubes, bronchitis, etc. It is better to know exactly . . . so, I am consumptive? Only since two or three years. And, in sum, it is not advanced enough to kill me, still it is very upsetting." [Marie Bashkirtseff, *Journal of Marie Bashkirtseff*, trans. Arthur D. Hall (Chicago: Rand, McNally, and Co., 1913), part 2, 220, trans. modified]. Note that it was in 1882 that Robert Koch identified the tubercular bacillus.

21. Cases where complacency in situations of illness aims to delay the inevitable return to a professional activity following medical leave are not what is in question here.

22. See Jean-Paul Velabrega, *La relation thérapeutique, malade et médecine* (Paris: Flammarion, 1962).

23. Michel Foucault, Blandine Barrett Kriegel, Anne Thalamy, François Beguin, and Bruno Fortier, *Les machines à guérir (aux origines de l'hôpital moderne)* (Paris: Institut de l'environnement, 1976).

24. Franz Kafka to Milena Jesenská, April 1920) in Franz Kafka, *Letters to Milena* (New York: Schocken, 1990), 6, trans. modified to align with Canguilhem's quotation. [For both this quote and the next, Canguilhem cites Klaus Wagenbach, *Kafka par lui-même* (Paris: Seuil, 1968), 137–38.—Trans.]

25. Franz Kafka to Max Brod, mid-September 1917, in Kafka, *Letters to Friends, Family, and Editors*, trans. Richard Winston and Clara Winston (New York: Schocken, 1977), 138.

26. See Professor P. Cornillot's reflections in "Quatre vérités sur la santé," in *Autrement 9, Francs-tireurs de la medécine* (1977). The author shows that the notion of absolute health is in contradiction with the dynamic proper to all biological systems and that as a result, relative health is a state of unstable dynamic equilibrium. "Relative health remains an apparent state, which implies no guarantee regarding the possible silent evolution of pathological processes that escape from the vigilance of natural mechanisms of struggle against aggression, infection, or depersonalization, in the biological or psychological sense of the term" (234).

In his *Histoire des expressions populaires relatives à l'anatomie*, Edouard Brissaud writes: "The most flourishing health does not presage the longest life. Try as one might to watch one's hygiene, to guard oneself from imprudence and vices that accelerate aging, in spite of everything, disease will still occur. Didn't one of our teachers—a hypochondriac, it is true—define health as a 'a precarious, transitory state that presages nothing good?'" Edouard Brissaud, *Histoire des expressions populaires relatives à la médecine* (Paris: Masson, 1892), 93–94. From this we may conclude that Dr. Knock was older than Jules Romains. [Trans.: *Knock, ou le triomphe de la médecine* was a popular play written by Jules Romains and first performed in 1923; Dr. Knock, its protagonist, takes advantage of townsfolk by offering free consultations and convincing them they suffer from various diseases that require long-term treatment.]

27. First published in 1934, Kurt Goldstein's *Der Aufbau des Organismus: Einführung in die Biologie unter besonderer Berücksichtigung der Erfahrungen am*

kranken Menschen (The Hague: Martinus Nijhoff, 1934) was translated into French under the title *La structure de l'organisme* (Paris: Gallimard, 1951). It is regrettable that to this day it has not been reprinted. [The French translation was republished in 1983; for the English, see Kurt Goldstein, *The Organism* (1938; New York: Zone Books, 1995). Canguilhem is referring to Merleau-Ponty's extended citations of Goldstein in *The Structure of Behavior* and *Phenomenology of Perception*. Canguilhem cited Goldstein more frequently than any other contemporary author; he also translated his work, and he was instrumental in the spread of Goldstein's thought in France. On the issue of normality and pathology, as well as that of the milieu, see Georges Canguilhem, *Knowledge of Life*, 113–14 and 129–132, and *The Normal and the Pathological* (New York: Zone Books, 1991), 181–96.—Trans.]

28. Goldstein, *The Organism*, 380.

29. [See Kurt Goldstein and Martin Scheerer, "Abstract and Concrete Behavior: An Experimental Study With Special Tests," in *Psychological Monographs* 53, no. 2 (1941):1–151; Kurt Goldstein, *Aftereffects of Brain Injuries in War*. (New York: Grune & Stratton, 1942).—Trans.]

30. Goldstein, *The Organism*, 341.

31. [Canguilhem is most likely referring to Michael Balint's influential *The Doctor, His Patient and the Illness* (London: Pitman, 1957).—Trans.]

32. Goldstein, *The Organism*, 341.

33. [See Ivan Illich, *Medical Nemesis: The Expropriation of Health* (New York: Random House, 1976).—Trans.]

34. See Evelyne Aziza-Shuster, *Le médecin de soi-même* (Paris: Presses Universitaires de Paris, 1972). [Trans.—For this theme, see also Chapter 1 of the present volume.]

35. [Canguilhem uses the French term for *petitio principii*—that is, for assuming the point one ostensibly proceeds to prove.—Trans.]

36. A great oncologist from Toulouse, justly respected for his generous devotion and indefatigable preoccupation with the personal problems of his patients, used to teach that stomach ulcers can be diagnosed over the phone.

37. [See Sigmund Freud, "A Metapsychological Supplement to the Theory of Dreams," in *The Standard Edition of the Complete Psychological Works of Sigmund Freud*, ed. and trans. James Strachey, vol. 14 (London: Hogarth Press, 1957), 222–35.—Trans.]

38. [F. Scott Fitzgerald, "The Crack-Up," in *The Crack-Up*, ed. Edmund Wilson (New York: New Directions, 1993), 69.—Trans.]

5. THE PROBLEM OF REGULATION IN THE ORGANISM AND IN SOCIETY

1. [Pierre-Maxime Schuhl (1902–1984) was a philosopher and historian of ancient Greek thought, and the author of *Essai sur la formation de la pensée grecque* (Paris: F. Alcan, 1934) and *Machinisme et philosophie* (Paris: F. Alcan, 1938).—Trans.]

2. Schuhl, *Essai sur la formation de la pensée grecque*, 307.

3. [Canguilhem addresses this problem at greater length in *Knowledge of Life* (New York: Fordham University Press, 2008), chapter 2, "Cell Theory." —Trans.]

4. Ernst Haeckel, *The Pedigree of Man and Other Essays*, trans. Edward B. Aveling (London: Freethought Publishing, 1883). See the section "Cell-Soul, Soul-Cells," where he describes "free republics of cells" and "cell monarchies," 170–72.

5. Gilbert Keith Chesterton, *What's Wrong with the World* (New York: Dodd, Mead, 1910).

6. [The regulator on a steam locomotive controls the amount of steam fed to the cylinders driving the wheels. It is the throttle for the steam engine. In the wake of the Second World War, "everyone" in Canguilhem's audience in 1955 would indeed be likely to know that a "regulating station" is a military command center established to control all movements of personnel and supplies into or out of a given area.—Trans.]

7. Walter Cannon first used the term "homoeostasis" in 1926 in "Physiological Regulation of Normal States: Some Tentative Postulates Concerning Biological Homeostatics," reprinted in L. L. Langley, ed., *Homeostasis: Origins of the Concept* (Stroudsburg: Dowden, Hutchinson, & Ross, 1973).

8. Claude Bernard. *An Introduction to the Study of Experimental Medicine*, trans. H. C. Greene (1865; New York: Dover Publications, 1957), 119.

9. [Canguilhem elaborates on these embryological experiments in *Knowledge of Life*, 14–15, specifically detailing the mineral, protein, and lipid elements of the ovarian cycle (in birds) and the embryo's morphological development, including the developmental limits of organic formation.—Trans.]

10. Walter B. Cannon, *The Wisdom of the Body*, 2nd ed. (1932; New York: Norton, 1963).

11. Henri Bergson, *The Two Sources of Morality and Religion*, trans. Cloudesley Brereton, R. Ashley Audra, and W. Horsfall Carter (London: Macmillan, 1935).

12. Cannon, *The Wisdom of the Body*, 311–12.

13. Ibid, 312.

14. Bergson, *The Two Sources of Morality and Religion*, 229.

15. Ibid., 252, in Bergson's final remarks, "Mechanics and Mysticism."

16. [Canguilhem is drawing from Cannon's discussion in the last section of *The Wisdom of the Body*, "Biological and Social Homeostasis."—Trans.]

17. Bergson, *The Two Sources of Morality and Religion*, 23–25.

forms of living

Stefanos Geroulanos and Todd Meyers, *series editors*

Georges Canguilhem, *Knowledge of Life*. Translated by Stefanos Geroulanos and Daniela Ginsburg, Introduction by Paola Marrati and Todd Meyers.

Henri Atlan, *Selected Writings: On Self-Organization, Philosophy, Bioethics, and Judaism*. Edited and with an Introduction by Stefanos Geroulanos and Todd Meyers.

Georges Canguilhem, *Writings on Medicine*. Translated and with an Introduction by Stefanos Geroulanos and Todd Meyers.

Jonathan Strauss, *Human Remains: Medicine, Death, and Desire in Nineteenth-Century Paris*.

Juan Manuel Garrido, *On Time, Being, and Hunger: Challenging the Traditional Way of Thinking Life*.

Milton Keynes UK
Ingram Content Group UK Ltd.
UKHW040843200224
438117UK00006B/217